V&R

Religion, Theologie und Naturwissenschaft/
Religion, Theology, and Natural Science

Herausgegeben von
Antje Jackelén, Gebhard Löhr, Ted Peters
und Nicolaas A. Rupke

Band 7

Vandenhoeck & Ruprecht

Ted Peters

Anticipating Omega

Science, Faith, and Our Ulimate Future

Foreword by Michael Welker

Vandenhoeck & Ruprecht

Mit 4 Abbildungen
und zahlreichen Grafiken

Bibliografische Information Der Deutschen Nationalbibliothek

Die Deutsche Nationalbibliothek verzeichnet diese Publikation in der Deutschen Nationalbibliografie; detaillierte bibliografische Daten sind im Internet über <http://dnb.dnb.de> abrufbar.

ISBN 10: 3-525-56978-5
ISBN 13: 978-3-525-56978-8

Printed in Germany.
Satz: OLD-Media OHG, Neckarsteinach.
Druck und Bindung: ⊕Hubert & Co, Göttingen.

Gedruckt auf alterungsbeständigem Papier.

Table of Contents

Foreword

This book offers eight essays on a new theology of hope. Forty years after Jürgen Moltmann's influential "Theology of Hope," forty years after Wolfhart Pannenberg's programmatic "Revelation as History," Ted Peters reconsiders central topics in eschatology. The innovative German theologians had engaged classical philosophies, biblical studies and theories of history for their enterprises. In his approach, Ted Peters seeks the discourse with scientists, particularly with those involved in the "science-and-theology" dialogue. A creative interaction between Christian theology and the various fields of the natural sciences leads to a deeper understanding of God's creativity in this world and of God's shaping of the world to come. Inspired by this dialogue and its many new insights into the nature of reality – earthly, human and divine –, Peters offers provocative news of God's "retroactivity": "God creates from the future, not the past!"

This retroactivity is illuminated with recourse to the resurrection of Christ and its meaning for a realistic hope, a realistic hope which asks for divine agency under the conditions of physical cosmology and biological anthropology. A passion to discover God's saving workings under the condition of this physical world and our biological existence also guides the following chapters which ask whether we can identify "Grace in the Genes" and discover an eschatological "Hermeneutics of Evolution." Two further chapters take up the challenges posed by the advances of neuro-biology, the cognitive sciences and the rhetorics and maneuvers of "posthumanism."

Peters does not seek easy answers to complicated questions and issues. He enters into discussion with a broad set of partners and problems, both epistemological and ethical. His goal is to give plausible arguments for the fact that we have to understand ourselves as biological beings and yet can sustain the belief that we are children of God, ennobled in ours lives in this world and for the world to come. A polyphonic yet consonant voice resonates in this book – the multidisciplinary academic and the apologetic representative of the Christian tradition, the innovative systematic theologian and the pastorally motivated minister help us to gain fresh insights and to affirm them, not only in truth- and salvation-seeking communities but also in environments skeptical over against eschatology.

Michael Welker

Preface

Faith gives birth to trust. Trust grows through relying on the God of our future. Trust matures as God proves trustworthy.

Trust in God provides the heart and blood for the Life of Beatitude, the blessèd life. For the person of faith, all the riches and wonders of our future life with God are present within the poverty and ordinariness of the present. From within the caverns of our own soul we hear first faint then louder echoes of God's call to trust in what is to come. In the very anticipation of the kingdom of God, somehow it is already here.

Like shaken champagne the Life of Beatitude fizzes beyond its original container, effervescing with humble self-acceptance; with heartfelt care for loved ones that moves us to mourning them at their passing; with hunger and thirst for justice and integrity; with magnanimity in showing mercy; with a passion for peace; and with purity of heart (Matthew 5:1–11). Masked beneath discouragement, we find passion for transformation. Masked beneath humility, we find greatness of vision. Masked beneath the present moment, we find the seeds of the future.

When we reflect intellectually on faith we pursue theology. Theology is the pursuit of expanded or deepened understanding. St. Anselm reminded us that faith inherently seeks understanding, *fides quaerens intellectum*. The task of this book, *Anticipating Omega*, is to pursue a theology that reflects on the God of promise who calls us to trust in our future, to trust in our ultimate future.

The axle around which everything spins is the Easter resurrection of Jesus and its proleptic connection – its identity – with the eschatological consummation of all God's work with the creation in the new creation. The raising of Jesus from the dead was impossible according to the laws of nature common to our world. Only the laws of a new world, a transformed world, could make such an event possible. The Scriptures promise that such a new world is coming. We find ourselves like a spark between two arcs, Jesus' Easter resurrection yesterday and our own resurrection tomorrow. The Life of Beatitude today is electrified by this in-betweenness.

Present reality is but a mask of future reality. In fact, I will argue in this book, God's future reality is our ultimate reality, the first reality. The present is but a shadow of the future. The future will so transform its past that we can say, reality is not done yet. Creation is not yet created. We are not yet who we are. Before we can know for certain what is and what has been, we will need to see what place it receives in what will be. Resurrection transforms all that has been prior; resurrection re-contextualizes it, re-defines it, re-essentializes it. The Life of Beatitude is like an oil well, it drills into the wealth of anticipation that lurks below the surface of the present.

In centuries past when theologians would engage in such reflection, they would partner with philosophy and draw a picture of the world in terms of metaphysics. In this book I will follow a parallel path; but I will partner rather with natural science. I will ask just how the eschatological vision of God's ultimate future fits on a map of reality along with physics, evolutionary biology, genetics, and technical prognostications for a post-human future. I will try to provide an ontology of the future, a retroactive ontology of the future. This will constitute an attempt to understand today's world of nature in light of tomorrow's world of transformation.

Although we have God to thank for our ultimate future, we have mortals to thank for making our daily life more edifying. I would like to thank Sheila Taylor for preparing the index to *Anticipating Omega*. And I would like to thank Tina Bruns of Vandenhoeck & Ruprecht for her conscientious and scrupulous copy editing. Both have contributed to the quality of this book's presentation.

Epiphany, 2006 *Ted Peters*

Chapter 1

Omega and Divine Retroactivity

Omega, Ω, is the final letter in the Greek alphabet. It marks not only the caboose on the train or the final page in the book, but it also symbolizes completion or fulfillment. The eternal God is our omega, just as God all along has been our alpha. Revelation 1:8 "'I am the Alpha and the Omega', says the Lord God, who is and who was and who is to come, the Almighty."

In light of this vision, I recommend that our ontology be based on Genesis 1:31, "God saw everything that he had made, and indeed, it was very good" along with Revelation 21:1, "Then I saw a new heaven and a new earth." Our epistemology should be based on I Corinthians 13:12, "now we see in a mirror dimly." Now, what does this mean?[1]

Omega and Retroactive Ontology

All things will be transformed at the omega coming of the Almighty. If this is our expectation, then just what is the ontological status of all things right now? Whatever things are, they will not remain as they are. Nothing remains the same. Nothing avoids undergoing change. We have long discerned that change involves deterioration, dissolution, death, and dissipation. Change may also incorporate temporary advancement in creativity, to be sure; but creativity will peak and then decline as well. So, process and change and even decline is a given. What can we expect to be the effect of omega?

Perhaps we can say that nothing in our present reality has attained its final state. Nothing has attained is final definition. Everything is subject to routine process and modification; and everything is subject to total redefinition at the coming of omega. Omega, then, becomes that which determines reality, that which defines who we are and what all things are.

[1] Scripture quotations are from the Revised Standard Version of the Bible unless otherwise stated, copyright © 1946, 1952, and 1971 the National Council of the Churches of Christ in the USA. Used by permission. All rights reserved.

Because God is gracious, according to the Christian claim, omega will be different from ordinary change. Ordinary change is associated with disease, death, and disappearance. Our gracious God promises healing, redemption, and renewal. We can expect omega to provide transformation, salvation, and eternal definition. The meaning of our existence and self-understanding today is contingent on our transformation in the future. Our final future will retroactively transform who we are today. It will determine who we had been as we anticipated who we would become.

I would like to call this line of thinking *retroactive ontology*. The fundamental insight is that our being is determined by, and defined by, our future. The transformed reality promised by God is the ground for all other reality that anticipates it. One important implication is this: creation is contingent on transformation. The meaning and even being of the past is contingent on its future. God's omega redefines – actually defines – all that has gone before. Who we are now is dependent on who we will be at omega.

In order to defend this constructive proposal and draw out some of its implications, I would like to set forth a series of theses. In the explication, I will at points refer to understandings of the natural world being discussed in the sciences. In particular, I will refer to Big Bang cosmology, evolutionary biology, genetics, and neuroscience. What each of these fields make clear is that nature is not static. Rather, nature is malleable, changing. Nature has a history; and it has a future. What we can see in nature is contingency; what we cannot see without looking through theological lenses is that the decisive contingency is God's promised action at omega. Connecting scientific perception with theological vision will be part of this agenda.

Thesis 1. God creates from the future, not the past.

To say that "God creates from the future, not from the past" may seem to fly in the face of common sense. As we look around at the objects in the room, we recognize that each one was manufactured. Should we investigate, we could find out where and when in time each item came into existence. Before that time, there was no such thing. Even the plants in the room came from seeds. Similarly, each one of was born once. Before our birth, we did not exist. So, common sense would place the creation of things in the past. What we see daily is the result of past creativity.

When we think of the cause and effect nexus of finite events, we place the cause in the past and the effect in the present. What exists in the present is the result of past causes. When we think of the creation of the world, then, we look to the past. We look to the *arche*, to the beginning, to the point of origin. We look backwards to alpha. What we find back at alpha we call *genesis*. Common sense places creation at genesis.

Now, let us look once again, a bit more closely. Just what does it mean to be, to have existence? Can we exist without a future? No. Without a future, we are not. If someone takes away our future, we drop from existence into non-existence. This is what death is, the loss of our future.

To be is to change. Being cannot be being unless it is becoming. Without giving the case entirely to Heraclitus, we must acknowledge that to be in the present is to be open toward multiple scenarios for the next moment as well as for the longer range future. An attempt to remain stationery or static leads to drowning and death in the never ceasing flow of temporality. Or, perhaps to reverse the ontological priority, being is time, *Sein ist Zeit*.

To be is to have a future. If this is the case, then the way God gives being is to give a future. Each moment, God gives the cosmos the next moment. God is moment by moment giving to all of realty its future. Without this future-giving on the part of God, all of reality would freeze up and cease. Romans 4:17b God "gives life to the dead and calls into existence the things that do not exist."

If we return for a moment to alpha, the moment our cosmos originated, perhaps we could say this: *the first thing God did for the cosmos was to give it a future*. By calling it from nothing into something, God bestowed a future that set reality on the course of historical becoming. By creation, we refer to God's gracious gift of futurity. And moment by moment with unceasing faithfulness God continues to bestow a future.

This dynamic relationship between God the future-giver and the creation which receives this gift includes another important element. That element is openness, openness to what is new. This means that God's future-giving is both positive and negative. It is positive in that God is the ground of being, the one who protects what is now from ceasing to be. God's work is negative, as well, in that by giving a new future God releases the present from the grip of the past. Contrary to common sense, past causes do not hold the present moment in the grip of absolute determinism. The present moment is open to change, open to what is new. This is because God liberates the present from the oppression of the past. New things can happen because God prevents the past from overpowering the present.

The first thing God did was provide nascent reality with an open future. Since then, God has continued this double relationship to the created order, negatively releasing the grip of the past while positively offering being and openness to a future of new possibilities. Yes, past causes may set the parameters that give specific form to the finitude of each present moment. Yet, within these parameters God opens up an array of potentials that await actualization. The way the creatures within the world behave determines which potentials become actualized.

In this fashion God's future-giving is what makes both contingency and freedom possible. The course of natural events is subject to contingency – that is, events in nature are not predictable. Despite the determinism of the laws of nature, natural history does not operate like a machine. New and unpredictable events happen. Then, when we turn to the course of human events, we take a giant step beyond contingency. Human actions are not predictable because they are freely determined. The human reality adds a subjective self that envisions the array of potentials and then makes a conscious decision to actualize some and not others. The openness of the future is the condition that makes freedom possible.

Still more needs to be said about God's future. We have just looked at one aspect of God's future-giving. Each moment God sustains the cosmos in being, provides an array of potentialities that makes contingency possible, and releases the present moment from the absolute grip of past determinism. This is God's proximate future-giving. In addition, God offers an ultimate future. The ultimate future is omega. God promises a final future, a fulfilling future, a moment when all reality will attain its full quidity. At omega, creation will be complete.

This distinction between the proximate and ultimate futures, as we look at it from the present moment, should not coincide with a distinction between the near future and the far future. The proximate and ultimate futures are not separated into a short time and a long time. Rather, both are almost present, almost but not quite fully here now. Both are as close to us as is the next moment. It is God's vision of the ultimate future that draws the present moment out of the nonbeing of the past toward the fullness of omega being, while to date allowing each moment to engage freely in self-definition and self-actualization. The ultimacy of God's destiny is only a moment away, practically present. In the twinkling of an eye, it could arrive in its fullness.

Thesis 2. God's creative action within nature and within history is derivative from the divine act of redeeming and creating the whole of the cosmos.

Creation is not a single act that happened just once a long time ago. Rather, creation is a single event incorporating the whole history of the cosmos, including but not limited to human history. Although we mistakenly think of God's creation as a once-upon-a-time event in the past, we unknowingly experience God's creative work as a constant durative process. Each moment of our lives God is drawing our personal reality out of the nonbeing of the past into the actual existence of the present moment, while maintaining a destiny for us, an as-yet-unrealized purpose which will become realized only in the consummate future.

Why did I speak of God's relation to cosmic history from alpha to omega as a single act? Let us try an idea on for size.

This idea is holism. The point of departure for holism is the principle that the whole is greater than the sum of the parts. A genuine whole is more than merely the sum total of aggregate items. A whole is an integrated system with an emergent property that is not reducible to the properties of any or all of the parts. This seems obviously true in the case of living organisms, and it is sometimes true in other systems. A living fish or a living human being cannot be disassembled into chemical components and retain the characteristics of a living organism. A whole in this organic sense is more than merely a mixture of chemicals or an aggregate of functioning parts.

Included in the holistic model for analyzing systems is the notion of downward causation. Whereas in upward causation the parts alter one another and the whole; in downward causation the whole alters the parts by incorporating their participation in the dynamics of the whole. For example, my heart is part of my body, part of me. Without my heart, I cannot be who I am. It pumps blood to keep me alive; and when it stops pumping blood, I will die. The whole of who I am is dependent on the heart. But, that is not all. I can influence what my heart does. If I decide to eat a large meal and follow it with a warm bath and a long night's sleep, my heart will lower its blood pumping rate; and I will get fat. If, as an alternative, I decide to grab a small snack and then go jogging, my heart will pump much faster, I'll feel energetic, and I'll lose weight. My decision as a whole person affects all my parts, including my heart.

Might we come up with a recipe for a holistic vision of creation? Can we mix together our ingredients thus far? Can we conceive of the creation as itself a whole, perhaps a whole of wholes plus parts? Can we think of God acting on the whole of creation and, thereby, reorienting and redefining all of the parts within? And if we add time to this recipe, can we predict the emergence of a future whole of wholes which will retroactively determine the place of all the parts? If we think of God's establishing point omega on analogy to baking a cake, can we think of God presently as mixing the ingredients? When the cake comes out of the oven, we could at that time consider all God's previous activity as the single act of baking a cake? All of God's current works are parts of a single comprehensive act of creating the world.

In the cake baking analogy, we see that time is important. While the ingredients were being mixed, no cake existed. The cake emerged as the final stage in the process. Only when the cake as a whole emerged could we say that all the parts contribute to the whole which is greater than the sum of those parts. Some wholes are emergent.

It is reasonable to consider the likelihood that our planet earth is a whole. The ecosphere is a complex system of interactions between living and nonliving components. Everything is in relation to everything else; and virtually all these relations need to be in tact for life to survive and thrive. Perhaps it is too much at this point to speak of the planet as itself a living organism named Gaia; yet, as earth's citizens, we feel some effects of downward causation.

The earth is by no means a closed system. It is an open system. Twenty-four hours per day our planet receives energy from the sun. The second law of thermodynamics describes how over time this energy input will lead to temporary periods of chaos followed by leaps into new and unpredictable order. Perhaps, in retrospect, this is why life evolved. Life is an emergent quality. The origin of life might be due to a natural advance in order as a result of energy input. Be that as it may, the point is that our third planet from the sun is not autonomous; it exists in relationship to our solar system, our galaxy, and to the cosmos.

Does this mean, then, that to think of God's creation we must think of it on a cosmic scale? Might holism apply on this scale?

At present, we cannot easily consider the entire cosmos as a single whole with downward causation. In various locations within the cosmos dynamic changes are taking place, super novas and similar change events that initially have little or no relation to what is happening in very distant regions of the same cosmos. Because light travels at a finite speed, and because the universe is so vast and galaxies are flying away from one another, the ripple effect of events in one galaxy simply cannot catch up to escaping galaxies in order to have an impact. What we mean by the word 'universe' might be a totality; but it does not look like a single whole.

Not only on the galactic scale but even on the human scale, many present events are contiguous with one another. They are parallel events, not links in a single causal chain. They do not affect each other, at least not immediately. Even if two events share the same causal history, they may not affect one another until some point in the future. They may eventually connect, to be sure.

Perhaps the concept of time can help us at this point. If we go backwards in time, say 13.7 billion years, we return to what appears to be a minute singularity. All things were but just one thing prior to the Big Bang. Although physicists disagree as to whether we can refer to the initial reality at the onset of the Big Bang as a singularity in the sense of an inert unit, we can say that all physical reality belonged together in a single inertial frame of reference. Only when the post-Big Bang expansion carried the developing universe beyond the threshold where all its light shared a common inertial frame did we lose the causal glue that holds the parts of the material world together.

Going back to the Big Bang incorporates time into our galactic story, to be sure; but this is historical time, not future time. Where we find ourselves in the present chapter of our universe's history is in the middle somewhere, and with a universe so strewn out that it may not even be a single universe any more.

What about the future? Physical cosmologists can only project on the basis of present observation that the universe will continue to expand for the foreseeable future. On the one hand, the second law of thermodynamics and its partner, the law of entropy, would suggest that as time passes and the universe expands further, it will die out. As we move from hot to cold, from concentrated energy to dissipated energy, like an old car the universe will run out of gas and just stop at a state of equilibrium. Particles will be so spread out and so cold that they simply will lose a causal relationship to one another. On the other hand, the observation that distant galaxies seem to be accelerating rather than slowing down leads to counter speculations. Might there be some sort of vacuum energy that accounts for this increase? Might this mean that entropy will not apply? In sum, based upon present observations, the future of our cosmos is barely understood or predicted.

The birdie that whispers in the theologians' ear tries to say: all things are but one creation by the one God of Israel. There is only one cosmos, and God is its creator and redeemer. God has but one plan and one destiny in mind for the future of all the galaxies, no matter how far apart they may be. At present, at least in the mind of God, all things belong together.

What remains to be asked is this: do all things including all galaxies constitute a single whole? Or does their wholeness exist only in the mind of God? And, if the latter, at what point will God's thought become physical reality?

At this juncture, perhaps we need to accept dissonance rather than consonance between the scientific vision of the far future and the eschatological vision. The winding down to a state of equilibrium due to entropy or other such scenarios that forecast the demise of physical reality do not dovetail conceptually with the biblical symbols of God's ultimate future. Isaiah 11:6 "The wolf shall live with the lamb, the leopard shall lie down with the kid, the calf and the lion and the fatling together, and a little child shall lead them." Or, Revelation 21:23 "And the city has no need of sun or moon to shine on it, for the glory of God is its light, and its lamp is the Lamb." Such biblical symbols of the new creation bespeak harmony, of a full integration of the divine life with cosmic life. They draw a holistic picture in which the contentious forces within the present creation are reconciled. Such salvation must apply to all things if it is to apply to one thing. If God be the creator of all that is, then nothing would be left out of this eschatological vision.

Such an eschatological future cannot result from the continuation of presently observable natural processes. The natural world would need transformation for the eschatological vision to become actual. This will have to be God's work. A future work of God such as this cannot be anticipated scientifically. It can be anticipated only by listening to the biblical promise.

Now to the central point of this thesis. This eschatological action by God will include the incorporation into the divine life of our cosmic reality. The creation will be absorbed into God, and God's presence will imbue the creation as a whole and in all its parts. God's love will become the glue that binds every sentient being into community with one another.

This applies to the temporality of creation as well. The entirety of past history will be taken up into eschatological eternity. The past will be resurrected and pass through judgment. Antecedents will become transformed by their consequences. Reconciliation will overcome division. Peace will overcome strife. Life will overcome death. The meaning of all past things will be established by their transformation. God's creative activity will attain its completion. God will be able to take that Sabbath rest described in Genesis 2:2 "And on the seventh day God finished the work that he had done, and he rested on the seventh day from all the work that he had done." That seventh day is tomorrow, the day that will conclude all of God's creative work. When it is redeemed, our world will be created.

Thesis 3. The concept of creation out of nothing (creatio ex nihilo) should be combined with continuous creation (creatio continua).
Although the opening verse of Genesis does not state unequivocally that God creates the material world out of nothing, this has become an indispensible assertion in Christian theology. *Creatio ex nihilo* is a conceptual explication of what theologians believe the biblical symbols are telling us. The point of this thesis is that *creatio continua* is an appropriate complement to *creatio ex nihilo*.

The logic of *creatio ex nihilo* draws in large part on implications of the gospel message of the Christian church. By 'gospel' I mean the story of Jesus told with its significance. Part of the significance is built into his Easter resurrection. As Jesus rose, God promises that we too shall rise. More. Romans 8:22 "We know that the whole creation has been groaning in labor pains until now;" and God promises to liberate creation from its travail. The new creation including our resurrection is the promise made to us by God when raising Jesus from the grave.

Now, we ask: what does it take to raise the dead? What does it take to consummate history into a new and everlasting kingdom? What does it take to make a new creation? It takes the same power that wrought the present creation in the first place. It takes mastery over the created order, total mastery. It takes a loving God whose power is undisputed and unrivaled. Only a God who can created something from nothing will be able to create life out of death.

For the Christian apologists of the first few centuries of the church, they found they had to deny the existence of pre-existing matter so as to avoid dualism. More than one eternal entity would compromise both. God could cotton no compromise with another eternal entity. Concluded the earliest systematic theologians: only God is eternal. The material world is temporal. It must have been made from nothing when it became temporal. The resurrection indirectly leads to positing this key understanding of creation, *creatio ex nihilo*.

This doctrine of creation defines God as creator. God's creative act begins with nothing; and then something is brought into existence. In the act of creating, God defines the Godself as Creator. Prior to the act of creating, God was God but not a creator. Once God initiated the creative process, the eternal God became related to time and space as its creator. Exactly how God should be defined as creator will depend on the nature of the created order, an order that is not done yet.

Now, there are two kinds of nothing. Nothing comes in a nondialectical and a dialectical form. Following Plato, nothing understood as *ouk on* or *nihil negativum* is nonbeing in the absolute or nondialectical sense. This is nonbeing as the abyss, nothingness as absence of everything. However, nothing as *me on* or *nihil privativum* is nonbeing in dialectical relation to being. This form of nothingness is the negation of what is. It is the power of destruction at work in the existing world. Dialectical nonbeing represents the quality of temporal passage that erodes what exists and leads to destruction, death, and dissolution.

Christian theologians readily assume that the concept of an original act of creation out of nothing must presuppose the nondialectical *ouk on*. Yet, I would not want us to overlook the role that *me on* can play in our concept of creation. God's dialectical relationship to *me on* proceeds over time. It warrants God's ongoing creative activity, which is also God's redeeming activity.

According to this thesis, we should describe God's abiding activity of continuing to provide the world with an open future in terms of continuing creation, *creatio continua*. Theologians of previous generations did not necessarily use this vocabulary. Presuming that the act of creation was a completed act in the distant past, theologians traditionally have employed other terms for God's sustaining of the universe and providential care. *Preservatio* or preservation refers to God's sustaining of existence. God preserves the world's existence by preventing it from dropping back into nonbeing. If we think of God's creation from nothing as the act prior to time that makes time and space possible, then we need a separate concept to identify God's activity within the temporal stream. *Conservatio* or concursus provides this. It refers to God's ongoing activities within time and space, concurrent with the activities of creatures such as ourselves. Concursus does not obviate the independence of creaturely activity; rather, it simply places God's activity within the world's temporal flow. *Gubernatio* or governance indicates that God's providence is not without purpose. God works with a promise or aim or goal. For God to achieve the divine aim we can expect further activity such as purifying, delivering, reconciling, and consummating the created order. Theologians working with such distinctions did not intend to sever the tie between God's creating and sustaining work, to be sure; in fact, the latter was seen as an extension of the former.

I would prefer to account for these traditional theological concerns under the heading of continuing creation. The act of drawing the world from nonbeing into being is not limited to a once-for-all event in the past. God is doing it right now. Without God's vigilant activity as what we call the primary cause, the world would immediately cease to exist and we would never know the difference. Moment by moment in a continuous temporal flow, God is drawing the world into being and protecting it from falling into nonbeing; and this accounts for what previous theologians thought to be God's sustaining or preserving work. Rather than preserving a creation already complete, I view God's creative work as ongoing; it is yet to become complete in the eschatological future. Right now, we creatures are on the way to becoming who we will be. So also is the entire creation still under construction, so to speak, yet to be completed and yet to be judged "very good."

Every moment, God draws the present forward in time, allowing the moment just past to slip off into nonbeing, into *me on*. The slipping of the present into the nonbeing of the just past is what releases the grip of the past from imprisoning us in a closed causal chain. The invoking of dialectical nonbeing is God's method for liberating each new moment for contingency and freedom. This is best called continuing creation, *creatio ex nihilo*. On the one hand, it looks somewhat like the concept of classical theists, according to which God sustains the world in being. On the other hand, it does not presume a static form of being. This theological vision relies more heavily on the notion of becoming, of reality as a durative temporal flow. Rather than sustaining something that already fully is, God is continually creating by offering what is a chance to become something new.

Thesis 4. God is the primary cause of nature's secondary causes.
I affirm as does virtually the entire Christian tradition that the God of Israel is the creator of all that is. In classical and especially medieval theology, God is identified as the primary cause of the world's existence; and the sequence of activities within the world are linked by secondary causes. From Aristotle on we have presumed that a secondary cause (*causa secunda*) precipitates a change or an effect in preexisting matter. It took God as the primary cause (*causa prima*) to bring matter into existence in the first place.

My own variant on designating God with the task of primary causation avoids dating the divine work of creation at a single moment in the past, say 13.7 billion years ago at the moment of the Big Bang. Rather, God's exercise of primary causation continues. It is exercised moment by moment or, better, as a ceaseless durative relation between creator and what is becoming created. The relations between creatures characterized by secondary causation is subject to scientific study. In our modern context, only the philosopher or the theologian can point in the direction of God as primary cause.

In classical theism, God's power is absolute. Yet, once the plan of creation is set in motion, God places absolute power in abeyance and ordains power to creatures to maintain the preordained plan of creation. God invests creatures with their own natural powers. Hence the medieval distinction (*potentia dei absoluta et ordinata*) between the absolute power of God and the power God ordains be exercised by the world's creatures.

This inherited classic view suggests two commitments, one to God's self-imposed constriction on the exercise of absolute power and the other to God's empowerment of the created order. What I would like to emphasize here is God's ordination or bestowal of the capacity for the world's own development, for evolution according to the world's own potential for self-organization. However, I hesitate to affirm what contemporary neo-orthodox theologians affirm, namely, that our omnipotent God engages in self-limitation, that God constricts divine power in order to make a power vacuum that can be filled with natural powers. The problem with this contemporary view is that it presumes a fixed pie of power, according to which God must take a smaller slice in order for the world to get a larger slice. In contrast to this view, I believe it is the exercise of God's power that empowers the world. God exercises this power duratively, faithfully maintaining the world in existence while granting partial release from the mechanistic grip of the past nexus of efficient causation. It is the exercise of God's power upon the world that makes contingency in nature and freedom for humanity possible.

The distinction between primary and secondary causation is an abstraction from the concrete flow of physical and extra-physical becoming. Primary causation is not absent when secondary is in effect; rather, they constitute two dimensions of a single reality. This permits a noninterventionist theory of divine action. God does not intervene in the sequence of secondary causes, because God's primary causation is always coincident with whatever is happening within the world. This does not preclude divine action at the secondary level; rather, it simply depicts God's ordinary relation to the world as that of primary cause while making no commitment to extraordinary interventions such as miracles.

Through the eyes of science what we see is the sequence of secondary causes. We do not see miracles, nor do we see primary causation. Science is free to limit itself to secular explanations for natural phenomena. Science provides theories of explanation within the self-imposed parameters of secondary relationships.

Thesis 5. Our human nature is not yet fully created; we are still becoming.

The human race in its entirety and each human person as an individual does not arrive complete as the result of God's primary creative action. Our creation is not something past. It is present. We are not done yet. We are not yet what God the creator intends for us to be. We are still baking, so to speak; we still need to emerge from the oven in our final form.

The human reality is a becoming reality. On a planetary scale, the human race is evolving and changing physically and culturally. On an individual scale, we are not born as mature adults. We grow and change and mature and deteriorate and die. At what point do we become human? Or, does our humanity include our entire evolutionary history and personal history? Does humanity require a biography? If so, then we are not yet there. We await a fullness yet to come.

Built into the definition of the human reality is life, eternal life. This is the way it was in the opening chapters of Genesis when the divine image was ascribed to the human race. When Jesus Christ – whom the New Testament describes as the true image of God, the *eikon tou theou* or *imago dei* – rose from the dead on Easter, this introduced resurrection into the definition of what a human being is. 1 Corinthians 15:22 "for as all die in Adam, so all will be made alive in Christ." We will not become who God intends us to be until we ourselves share in the resurrection at omega. Once raised, we will look back over our biographies and over the evolutionary biography of the entire human race and understand who we are in our totality. Who we are will be defined by the length and breadth of our relationship with our Creator God.

As we examine the human condition as we find it now, we are examining a slice of human history and only a slice of the human reality. We can incorporate some of our past – both our prehistoric and historic past – but we cannot except in only the most vague way incorporate our future destiny. We must define our human reality as best we can by including some speculation about our ultimate future.

Thesis 6. Our human reality is in continuity with the surrounding natural world, including suffering and sin.

DNA is DNA is DNA. The four chemicals ACGT that make up DNA are the same in all life forms on planet earth. Whether with chimpanzees, Chihuahuas, chickens, Chinooks, or chickpeas, we human beings share a pattern of genetic activity. Just as Genesis 2:7 says we are formed from the soil, *adamah*, and therefore are made up of our planet's material, we can also say that what makes us live is in continuity with what makes anything alive in our world. And if Darwin's theory of evolution holds, we will have to conclude that all living things share a single biological history, perhaps even a single biological origin.

The relational model for understanding the human condition warrants adherence. The relationalist model cedes ontological priority to the relationship and subordinate status to the relation's terms. What we know as human individualism finds its place within the more comprehensive network of relationships. The human soul, for example, is not an isolated metaphysical monad, but rather a cipher that locates the dimension of the human reality that overlaps with the presence of God in a person's life. God has a relationship with each of us individually, to be sure; yet, even this relationship is tied to the network of connections with the rest of reality. Our resurrection, to cite a second example, is tied to the advent of the new creation, the eschatological renewal of the entire created order.

When it comes to understanding the shadow side of human existence – evil, frustration, suffering, sin, and death – we need to look first at the continuity between the human race and other forms of life, even preanimate matter. We need to look at nonhuman life forms to compare our experience of suffering with that of other animals. We need to ask whether our evolutionary history has disposed human beings for violent and destructive behavior, and whether even sinful behavior might be commonplace in the wild kingdom. We might even want to ask if preanimate physical matter operates according to principles that make possible the rivalry for resources and the conflict between life forms that leads to the predator and prey relationship. The theological concept of the fallenness of nature needs to be explored in light of a relationalist model for apprehending the interconnectedness of the living and nonliving domains.

Thesis 7. We need to think epigenetically, not archonically.
When trying to get at the basic structure or even essence of something such as the cosmos or significant items within the cosmos, two paths are available: the archonic and the epigenetic forms of thinking. The *archonic* path takes us back to the origin, to the beginning. I elect this term because its Greek root, *arche*, has a relevant double meaning. It means both beginning and governance. In the word *archaeology* it means origin, for example; and in *monarchy* or *hierarchy* it means governance.

The word *arche* betrays a structural propensity in human thinking, namely, we associate the definition or essence of something with its origin. The rule or governing principle is coincident with the way something begins. The nature of things is established at origin, so whatever comes subsequently is a betrayal or deviation from a thing's original nature. Analysis takes the form of seeking origins, because in an origin we believe we find the essence. What this leads to is the concept of revolution – to revolve – as a return to an original essence by means of clearing away unwanted accruals.

Myth provides the paradigm of archonic thinking. Here is my working definition: a myth is a story about how the gods created the world in the beginning, *in illo tempore* or the time before there was any time, which explains why things are the way they are today. Whether providing etiologies for the cosmos or one's nation or name or disease or whatever, the myth reveals a thing's essential nature by telling us a story of its origin. We may tell very few myths in the modern world, but the archonic path is still followed. Physical cosmologists still look back to the Big Bang in hopes of devising a Grand Unified Theory (GUT). Medical researchers still look for the ontogenesis of maladies. Archonic thinking is common to both myth and science.

The other path is *epigenesis.* If we take this word apart, we can see that 'genesis', reminding us of 'generate', refers to bringing something into existence for the first time. It is no accident that the First Book of Moses or the Torah is called "Genesis." If we prefix it with 'epi', the Greek preposition for 'upon' or 'after', we get a compound word that suggests ongoing or repeated genesis. Epigenesis is the process by which new things continue to emerge. The way the world begins is not the way it remains. It changes. New things appear. New things do not need to bow in allegiance to what preceded them, nor is their essence reduced to the nature of their predecessors. The reality of new developements is not suspect because they were not present at the point of origin.

What does this imply for theology? To be sure, much of the Bible follows the archonic path. The book of Genesis is rife with etiological narratives, stories about the origin of virtually every aspect of daily life for Israelites in ancient Canaan. The origin of Israel is given in the story of Abraham's call; and the name 'Israel' is given in the story of Jacob. Yet, this is not all. The God of Israel makes promises to do new things; and this God fulfills these promises. The people of Israel live between promise and fulfillment. God is not stuck in the past. God's future relationship to the covenant people is not determined by what happened at origin. Isaiah 43: 19 "I am about to do a new thing; now it springs forth, do you not perceive it?" We will miss perceiving God's new things if our eyes look only toward the past and not toward the future.

The path of epigenesis helps us perceive the dynamism of every day existence as well as view with credibility the divine promise for an eschatological consummation where God will become all in all. New things happen every day, both in natural history and human history. The eschatological transformation is continuous with daily newness, even though the breadth of the eschatological transformation is total. Only by liberating our thought processes from reliance upon archonic ontologies can we come to appreciate the Christian gospel that anticipates omega.

Thesis 8: Genesis can be interpreted eschatologically.
We have observed above that new wholes transform past parts. Integration into new more comprehensive unities preserve while renewing what came before. This holistic complexification process is nonlinear. Adding a new whole changes an entire situation in a significant way. The possibility of transformative effect renders redemption possible. Now, suppose we apply this to eschatology and then to creation? God's eschatological redemption will so reconfigure all that had been past that it might as well be a new creation or, perhaps more accurately, the completion of the creation already begun. Does this mean that eschatological omega takes ontological priority over what happened at the beginning? I believe it does.

The new heaven and new earth prophesied by the biblical Apocalypse will transform yet preserve the entire history of cosmic creation. What God did at the beginning to draw the physical world from nonbeing into being along with God's continuous sustaining of the natural order during its period of self-organization will be taken up into the consummate new creation. God's creative activity within nature and within human history is derivative from his eschatological act of redeeming the whole of the cosmos. Where we find ourselves today is looking back to alpha, to *creatio ex nihilo*, and looking forward to omega, the new creation *ex vetere,* out of what has come before.

The new creation is what is described in the earliest chapters of Genesis, in the Garden of Eden. Revelation 21:4 God "will wipe every tear from their eyes. Death will be no more; mourning and crying and pain will be no more, for the first things have passed away." The violence, suffering, and death so inescapable in today's world will become only a past memory. This is the component of redemption in the new creation.

Obvious here is that the arrival of the new heaven and new earth will not signal the total abandonment of the old heaven and old earth. The new creation will consist of refashioning the present creation, a weaving of the spiritual thread throughout the fabric of physical reality. God's grace will clothe every creature, and every creature will freely don its redemptive garments. God's spirit will imbue everything bodily while everything bodily will retain an enhanced self-identity. The creation will constitute a single all encompassing sacrament. Now, I ask: how does this eschatological vision influence our interpretation of the Bible's opening, Genesis 1:1–2:4a?

In the beginning, God created by uttering the Word. Actually, God's Word is not limited to the beginning. God's Word is eternal; and it is incarnate in Jesus Christ. And we receive God's Word in the form of promise. God plans to keep his Word.

Now, might we think of the opening account of creation in Genesis 1:1–2:4a as describing not merely a beginning but also the present time? Could we think of the creation week of seven days as inclusive of the entire history of the creation from Big Bang to whatever will become of the universe in the future? Could evolutionary history constitute one small episode in the divine epic of creation? Note how God asks the earth to bring forth living things. Genesis 1:24 "God said, 'Let the earth bring forth living creatures of every kind: cattle and creeping things and wild animals of the earth of every kind'." Could this mean that God the primary cause ordained earth as secondary cause to evolve? Could we see this as a process that still envelopes us? Could we see ourselves today standing between the initial moment when God opened his divine mouth to say, "Let there be ..." and the final moment when God declares that, behold, it is "very good"? Could we still be looking forward to the Sabbath day, to God's first day of rest yet in the future?

Thesis 9: The Life of Beatitude *daily anticipates omega.*

What I mean here by anticipation is pre-actualization of a still outstanding future. It is the equivalent of incarnation, incarnating in the present in our person what is yet to come at omega. I use the word *prolepsis* to emphasize the ontological weight of anticipation. A prolepsis is an embodied anticipation.

The paradigmatic model for proleptic thinking is the relationship between the eschatological resurrection and Jesus' resurrection. What happened to Jesus on the first Easter was a prolepsis of the new creation, an anticipation of the final resurrection that will include you and me. 1 Corinthians 15:20 "Christ has been raised from the dead, the first fruits of those who have died." As Jesus rose on Easter, so also will we rise at omega.

There is more to the Christian claim. Now, at the present moment, that new life promised for omega can have an affect. The resurrected Christ does not simply await us; new life can imbue our present life. Romans 6:3–5 "Do you not know that all of us who have been baptized into Christ Jesus were baptized into his death? Therefore we have been buried with him by baptism into death, so that, just as Christ was raised from the dead by the glory of the Father, so we too might walk in newness of life. For if we have been united with him in a death like his, we will certainly be united with him in a resurrection like his."

This walking daily in new life I refer to as the *Life of Beatitude*. The proleptic structure is almost visible in Jesus' Beatitudes in the Sermon on the Mount. The future kingdom of God seems to affect if not imbue the kind of person Jesus describes in his aphorisms. Matthew 5:5–9: "Blessed are the meek, for they will inherit the earth," says Jesus. "Blessed are those who hunger and thirst for righteousness, for they will be filled." Or, "Blessed are the merciful, for they will receive mercy." Or, "Blessed are the pure in heart, for they will see God." And, "Blessed are the peacemakers, for they will be called children of God." The dialectic of future with present is vivid. The anticipated kingdom of God is present – almost unconsciously present – in we who live the life of beatitude; and our blessing will be revealed at the advent of God's final kingdom.

Conclusion

With these theses I have tried to lay out a schematic map for an integrated set of theological concepts. When we zoom and re-center our computer generated map, we notice lines indicating streets running to the edge of the map. Those lines go somewhere, many places. So also, this schematic map of theses sends roads and paths in many directions. We cannot follow them all here. In the chapters that follow, we will zoom out and follow a few selected lines that take us in the direction of creative interaction between Christian theology and the natural sciences.

Chapter 2

Resurrection: The Prolepsis of New Creation

Like the binding on a book that holds all the pages together so they can be viewed individually, the concept of prolepsis holds together the various parts of this theological system. It is prolepsis that connects the history of Jesus that first gave rise to the Christian faith with the speculative ontology anticipating God's promised omega for all things. All things gain their being and their meaning, not from their moment of origin, but from their place in the new creation. What we are will be determined by what we will become. To anticipate the determining of our essence by what we will become is to think proleptically. To live daily out of this vision is to live beatitudinally.

We live in the era of the church. The Christian church is an arc between two terminals, a connection between the Easter resurrection of Jesus Christ yesterday and the advent of the kingdom of God tomorrow. The challenge is to keep the spark alive. The spark of the arc brightens our life when we live daily out of God's promise for the future.

The construction of the concept of prolepsis requires a double hermeneutic. On the one hand, we review the witness of scripture through the eyes of the exegete to ascertain the historical basis for the Christian claim. Theology is the explication of fundamental symbols, and those symbols are found primarily in the Bible.

On the other hand, we do not remain at the stage of exegesis. We proceed toward worldview construction, toward painting a picture of the whole of reality oriented toward God's promise of a new creation. To paint this picture in our modern and emerging postmodern context, our brush must dip into the colors provided by natural science. Not only natural science, to be sure. Other fields expand our horizon and enhance our insight as well. But everything cannot be tackled at once. In what follows, we will focus on relevant aspects of our physical reality that are affected by theological anticipations of omega.

Here is the central question of this chapter: how should we think historically about the Easter resurrection of Jesus, and how should we think theologically about the eschatological promise of our resurrection into the new creation? My answer has to do with prolepsis, with seeing Jesus' Easter resurrection as an anticipation or prefiguration of what God promises for us. "The resurrection functions as a promise that creation has a future in God," writes Australian theologian Denis Edwards.[1]

As we explicate this promise theologically, we cannot help but ask: just what might resurrected existence look like? Might it look like life as we know it now, perhaps tweaked here and there with immortality? Or, might it require such a radical transformation that it defies our imagination? Or, might it lie somewhere in the middle?

To address such questions in the contemporary context within which we find ourselves, the resources of modern science are ready-to-hand. How might we imagine a resurrected reality in light of what we know about the present reality from the field of physics? How radical a change would we have to expect?

As we proceed to develop an answer, I will draw on the scholarship of biblical historian, N.T. Wright, systematic theologian Wolfhart Pannenberg, and two scholars who reflect theologically on the implications of physics, John Polkinghorne and Robert John Russell. The physicists agree that the Christian concept of resurrection is not simply one more form of belief in immortality. Rather, it implies a transformation of the entire natural universe. It implies a transformation of the very principles by which the physical universe is governed. It implies a new creation, anticipated by the Easter resurrection. Might we think of the Easter resurrection of Jesus the way Robert John Russell does, as the first instantiation of a new law of nature?

Was Jesus' Resurrection Historical?

The debate over the resurrection of Jesus raises one of most important issues in the understanding of reality, says Christoph Schwöbel.[2] In the construction of a picture of reality in relation to God, it is important to proleptic theology to view the Easter resurrection of Jesus as historical. Why? Because the significance of Jesus has implications for objective reality, for history, and even for nature.

---

[1] Denis Edwards, *Breath of Life: A Theology of the Creator Spirit* (Maryknoll NY: Orbis Books, 2004) 110. To say that the resurrection of Jesus is prototypical of our resurrection does not exhaust the significance of what happened at Jesus' first Easter. Jesus' resurrection bestows salvation; whereas our resurrection marks our receiving of salvation. See: Gerald O'Collins, *Jesus Risen: An Historical, Fundamental and Systematic Examination of Christ's Resurrection* (New York: Paulist Press, 1987) 180.

[2] Christoph Schwöbel, "II. Auferstehung Jesu Christ; 2. Dogmatisch," in *Religion in Geschichte und Gegenwart*, ed. by Hans Dieter Betz/Don S. Browning/Bernd Janowski/Eberhard Jüngel (8 Volumes: Tübingen: Mohr Siebeck, [4]1998–2005) 1:926 (Hereinafter: RGG[4]).

Proleptic theology cannot be satisfied with strictly subjectivist approaches to resurrection, such as one might find in the Bultmannian tradition. That Jesus Christ rose into the faith of the messiah's disciples is doubtlessly essential to understanding the birth of the early church. Yet, the subjective appropriation is not enough. I'm sufficiently persuaded by the Second Quest for the Historical Jesus to postulate that an objective historical event that matches the first faith must have prompted that first faith. Even if the details of this event are difficult to retrieve due to layers of subsequent interpretation, it is important to presuppose if not demonstrate that Jesus rose from the dead and was witnessed in his resurrected state. "This reality of the resurrection is no illusion, no product of the fantasy," writes Michael Welker. "It is borne witness to by many men and women, and in many different contexts of experience."[3]

On this matter the exegesis of British biblical scholar, N.T. Wright, provides direction. He asks: why did the early Christian church come into existence? Wright answers: "the *only* possible reason why early Christianity began and took the shape it did is that the tomb really was empty and that people really did meet Jesus, alive again ... the best historical explanation for all these phenomena is that Jesus was indeed bodily raised from the dead."[4] The Easter resurrection of Jesus gave birth to the church. "The combination of empty tomb and appearances of the living Jesus forms a set of circumstances which is itself *both necessary and sufficient* for the rise of early Christian belief. Without these phenomena, we cannot explain why this belief came into existence, and took the shape it did. With them, we can explain it exactly and precisely."[5]

Wright's task here is that of the historian, not that of the theologian, at least not directly. Wright's argument proceeds as follows: (1) from the very earliest days Christians shared stories of both the empty tomb and of appearances of the resurrected Jesus, and those early oral stories were very similar to the ones we now have in written form in our Gospels; (2) Paul, far from offering a different, more spiritual view of Jesus' resurrection, is actually summarizing those early stories in the opening verses of 1 Corinthians 15 and elsewhere, and his great contribution was to provide, not a different story, but a theological way of thinking about those traditional stories, already well known in his day. "The gospel stories are not dependent on Paul ... Irrespective of when

---

[3] Michael Welker, "Resurrection and Eternal Life: The Canonic Memory of the Resurrected Christ, His Reality, and His Glory," in *The End of the World and The Ends of God: Science and Theology on Eschatology*, ed. by John Polkinghorne/Michael Welker (Harrisburg PA: Trinity Press International, 2000) 283.

[4] N.T. Wright, *The Resurrection of the Son of God*, Volume III of *Christian Origins and the Question of God* (Minneapolis: Fortress Press, 2003) 8, Wright's italics.

[5] Ibid., 696, Wright's italics.

the gospels reached their final form, the strong probability is that the Easter stories they contain go back to genuinely early oral tradition."[6] In summary, Wright contends that the empty tomb combined with the appearances created the church.

As suggested above, we need to ask: was the Easter resurrection a historical event? Many say "no," either skeptically because dead people simply don't rise or faithfully because faith is said to be subjective and not objective. The skeptics argue against historicity on the grounds of analogy: because dead people normally remain dead, Jesus must have remained dead as well. Ernst Troeltsch bequeathed to modern scholarship a formalized principle of analogy – that is, because he has himself not observed a dead person rising he argues, by analogy, neither Jesus nor anybody else could possibly have risen in the past.

Wright does not belong in the skeptic's camp; nor does he allow the principle of analogy to get in the way of pursuing historical study. He concedes that Jesus' Easter resurrection was unique, that it does not have an analogy in common experience. "*The fact that dead people do not ordinarily rise is itself part of early Christian belief*, not an objection to it. The early Christians insisted that what had happened to Jesus was precisely something new; was, indeed, the start of a whole new mode of existence, a new creation."[7] The uniqueness of what happened to Jesus is part of its significance. "It is precisely the uniqueness of the rise of the early church that forces us to say: never mind analogies, what happened?"[8]

Wright does not belong in the fideist camp with Rudolph Bultmann and existentialist interpreters. The resurrection of Jesus cannot be reduced to the rise of faith in the disciples of Jesus, or in the renewal of hope among the members of what would become the Christian church. A subjectivist explanation for the appearance of the church is insufficient. The followers of Jesus who began the church experienced a new reality, an objective reality, to which the rise of faith became a response.

From Past to Prolepsis

The Easter resurrection of Jesus is a past event, an event available for historical study. Yet, Easter could not be what it was without the accompanying promise of the eschatological kingdom of God. The past

---

[6] Ibid., 612.

[7] Ibid., 712, Wright's italics.

[8] Ibid., 18. For a thorough critique of the Troeltschian overuse of the principle of analogy in historical research, see: Wolfhart Pannenberg, *Basic Questions in Theology* (2 Volumes: Minneapolis: Fortress Press, 1970–1971) I:38–53.

cannot be what it is without its future. The anticipation of omega is built right into the meaning of the original Easter event.

Wright sees accurately that what happened to Jesus can be understood only within the framework of God's promise of our future resurrection into the new creation. "Early Christianity was a 'resurrection' movement through and through, and that, indeed, it stated much more precisely what exactly 'resurrection' involved (it meant going through death and out into a new kind of bodily existence beyond, and it was happening in two stages, with Jesus first and everyone else later); second, that though the literal 'resurrection' of which the early Christians spoke remained firmly in the future, it colored and gave shape to present Christian living as well."[9] This is the proleptic structure: Jesus' Easter resurrection is inextricably linked to the promise of our resurrection.

Wright employs the term 'prolepsis', but not extensively. Yet, it is the theological concept that connects the basic historical experience with Jesus to the eschatological significance or meaning of Jesus. "The heart and centre of it all, then, is the defeat of death in the future, based on the proleptic defeat inflicted in the resurrection of Jesus himself; or, to put it another way, it is the final completion of the 'age to come', which was inaugurated, in the midst of the 'present evil age', through the Messiah's death and resurrection."[10] With prolepsis, we understand the Easter resurrection of Jesus as an anticipation or even a prefiguration of the advent of the eschatological new creation. This is a theological judgment, already present in the New Testament. It is built right into the meaning of the history of Jesus.

From History to Meaning

We need to take a step from historical judgment toward theological meaning, from reference toward significance. Where we intend to go is from historical exegesis to systematic theology. Jürgen Moltmann provides the systematician's vision of proleptic eschatology. "If Christ has been raised *from* the dead, then he takes on the proleptic and representative significance *for* all the dead."[11] The connection between Jesus' Easter resurrection and God's promised future looks like this: "with the raising of the crucified Christ from the dead, the future of the new creation of all things has already begun in the midst of this

---

[9] Wright, *Resurrection,* 210.

[10] Ibid., 336.

[11] Jürgen Moltmann, *The Coming of God: Christian Eschatology,* tr. by Margaret Kohl (Minneapolis: Fortress Press, 1996) 29, Moltmann's italics.

dying and transitory world."[12] Now, we must ask: just how is this conceptual meaning drawn from our historical interpretation? It takes the form of drawing implications out of what the biblical text says.

N.T. Wright acknowledges the hermeneutical distinction between referent and meaning. By 'referent' he refers to the historical judgment that Jesus rose bodily from the dead. By 'meaning' he draws out its significance, namely, Jesus is the Son of God. "The resurrection, in other words, declares that Jesus really is God's Son: not only in the sense that he is the Messiah, though Paul certainly intends that here, not only in the sense that he is the world's true lord, though Paul intends that too, but also in the sense that he is the one in whom the living God, Israel's God, has become personally present in the world, has become one of the human creatures that were made from the beginning in the image of this same God."[13]

As we move from exegesis to systematic theology, this vision of a proleptic connection between the historical Easter and the eschatological promise is our focus. Yet, we need to ask whether the systematic theologian finds the connection already built into the history, or must the connection between resurrection and new creation result from a constructive interpretation of what Scripture says. Or, a combination of both?

The systematic theologian needs to rely on a two directional hermeneutic. The first direction is backward, back from the biblical text toward speculation about the history that must have given rise to the text as we read it. The second direction is forward toward our contemporary context, toward an explication of the Christian faith within the modern post-Enlightenment horizon of meaning. Both directions start from the same location, from what scripture says. Both history and explication are interpretations of scripture. The horizon of the future plays a significant role in both.

From Meaning to Ontology

The Easter resurrection of Jesus was more than a nice thing that happened to this one person one day in Jerusalem. Built right into the meaning of the Easter event is God's promise to establish the divine kingdom, to renew all of creation. This promise becomes the window through

---

[12] Ibid., 136.

[13] Wright, *Resurrection*, 733. See my chapter, "The Future of Resurrection" in *The Resurrection: John Dominic Crossan and N. T. Wright in Dialogue*, ed. by Robert Stewart (Minneapolis: Fortress, 2006) 149–170.

which we view ultimate reality. The eschatological new creation, omega, will redefine all of nature and all of history, all that has preceded it. The story of Jesus' resurrection leads the theologian step by step to hypothesizing a retroactive ontology, according to which the anticipated omega will so reorient the history of reality as to redefine it.

To take the next step in developing the proleptic significance of Jesus' resurrection and the accompanying promise of our resurrection, we can draw on the scholarship of Munich systematic theologian Wolfhart Pannenberg. He emphasizes that Jesus' Easter resurrection was a pre-realization of God's eschatological promise. "The ministry of Jesus shared with prophecy the character of an anticipation, but not only as in prophecy and apocalyptic in the sense of mere pre-*cognition*, but, so to say, as a pre-*realization* of the future, as its proleptic dawning."[14]

Actually, according to Pannenberg, we do not first find bare history and then add significance on top of it. Rather, history and significance come together in a single package. Wrapped right within the original meaning of Jesus' Easter resurrection is the dawning of the general or universal resurrection previously anticipated in apocalyptic prophecies. Whether Jewish listeners to Jesus believed or disbelieved in the apocalyptic vision of resurrection, they understood it; and they could understand why what happened to Jesus on Easter would be interpreted within this horizon. Jesus' resurrection is the "first fruits" of those having "fallen asleep" (1 Cor. 15:20). What was expected for all of us has already happened individually to the person of Jesus. "Jesus' expectation of the speedy realization of the eschatological reality did not simply fail. It was fulfilled, and thus confirmed, though only in his own person ... The general human destiny has occurred in Jesus – if he *really was* resurrected from the dead."[15]

It is important to note how Pannenberg makes his theological assertion contingent upon a historical one: "If he *really* was resurrected from the dead." Theological commitment is dependent upon historicity of the founding event, Jesus' personal resurrection. Faith is dependent on history. Faith, even though it is a subjective appropriation of Christian belief, is not reducible to its subjective character; it requires grounding in the objective realm of fact. This means, among other things, that a historical judgment against the authenticity of the biblical claim regarding Jesus' resurrection would undermine Christian belief.

---

[14] Wolfhart Pannenberg, "Focal Essay: The Revelation of God in Jesus of Nazareth," in *Theology As History*, Volume III of *New Frontiers in Theology*, ed. by James M. Robinson/John B. Cobb, Jr. (New York: Harper, 1967) 112–113, Pannenberg's italics.

[15] Ibid., 114, Pannenberg's italics.

The implication of Pannenberg's method is that we need a more thorough integration of fact and meaning, what Wright calls 'referent' and 'meaning'. "Such splitting up of historical consciousness into a detection of facts and an evaluation of them ... is intolerable to Christian faith, not only because the message of the resurrection of Jesus and of God's revelation in him necessarily becomes merely subjective interpretation, but also because it is the reflection of an outmoded and questionable historical method. It is based on the futile aim of the positivist historians to ascertain bare facts without meaning in history ... Against this we must reinstate today the original unity of facts and their meaning."[16]

Just how close are Wright and Pannenberg on this matter? Is Wright more willing to separate historical fact from its meaning? Is Wright insistent that we first get the historical facts straight, then we proceed to make theological judgments? Or, would it be better to see fact and meaning inextricably united in the historical reports of Jesus' resurrection and then consider our division of fact from meaning to be an abstraction? Pannenberg tips toward the latter.

Theologians such as Pannenberg proceed with the assumption that historical reconstrual of the facts will confirm the historicity of the Easter resurrection. He proceeds accordingly. "Now let us assume that the historical question concerning the resurrection of Jesus has been decided positively, then the meaning of this event as God's final revelation is not something that must be added to it; rather, this is the original meaning inherent in that event within its own context of history and tradition."[17]

As we explicate the meaning of Jesus' Easter resurrection, we draw out its significance for Jesus' person and for the eschatological vision. According to Pannenberg: If Jesus has been raised, then (a) the end of the world has begun; (b) God himself has confirmed the pre-Easter activity of Jesus; (c) Jesus is the expected Son of Man; (d) God is ultimately revealed in Jesus; (e) the universality of Israel's God motivates the gentile mission; (f) Jesus' words and appearances should be interpreted as having the same content.[18] The methodological point is that for both the early oral and written tradition as well as people of Christian faith today, the historical report of Jesus' Easter resurrection comes prepackaged, so to speak, wrapped in the eschatological promise of new creation and the advent of God's everlasting kingdom. Both this refer-

---

[16] Ibid., 126–127.

[17] Ibid., 128.

[18] Wolfhart Pannenberg, *Jesus – God and Man*, tr. by Lewis L. Wilkins/Duane A. Priebe (Louisville: Westminster John Knox Press Press, ^{2}1977) 66–73.

ent and the meaning of this historical event are contingent upon their eschatological confirmation or disconfirmation.

The inextricable connection between past and future, between what happened to Jesus and what will happen to us and the whole of reality, is built into the historical question. Adolf Harnack recognized this in the development of dogma. "The firm confidence of the disciples in Jesus was rooted in the belief that he did not abide in death, but was raised by God ... Consequently, the resurrection of Jesus became the sure pledge of the resurrection of all believers, that is of their real personal resurrection."[19]

Pannenberg explicates it systematically by arguing that the proleptic events surrounding the historical Jesus and the eschatological event of new creation are a single reality, a single act by an eternal God with a temporal world. "The coming again of Christ will be the completion of the work of the Spirit that began in the incarnation and with the resurrection of Jesus. From the standpoint of eternity we have here one and the same event because the incarnation is already the inbreaking of the future of God, the entry of eternity into time. For us, however, confession of the incarnation has its basis in Jesus' resurrection, and only at his return will debate concerning the reality of the Easter event be at an end and will that reality definitively and publicly come into force, for the resurrection of Jesus is a proleptic manifestation of the reality of the new, eschatological life of salvation in Jesus himself."[20] Only in the future will we know the past for certain.

Eschatology and Physical Cosmology

As previously mentioned, the systematic theologian needs a two directional hermeneutic. Beginning with scripture, the theologian goes backward to speculative history and forward to explications of the biblical faith. As we move forward from the biblical horizon to the horizon of understanding of the modern world, we encounter modern science. Science influences if not determines the conceptual apparatus of the contemporary mind. Proleptic thinking needs science. Within this context we might ask: just how can we conceive of the resurrection

[19] Adolf Harnack, *History of Dogma*, tr. by Neil Buchanan (7 Volumes: New York: Dover, 1961) I:84–85.

[20] Wolfhart Pannenberg, *Systematic Theology*, tr. by Geoffrey W. Bromily, (3 Volumes: Grand Rapids MI: Wm. B. Eerdmans, 1991–1998) III:627. See: Ted Peters, "Clarity of the Part versus Meaning of the Whole," in *Beginning with the End: God, Science, and Wolfhart Pannenberg*, ed. by Carol Rausch Albright/Joel Haugen (Chicago: Open Court, 1997) 289–302.

of the body? Any answer we offer will, like history, be speculative and, thereby, hypothetical and provisional. But explication requires such theological pre-construction.

Perhaps at this juncture we should turn our attention to the creative contributions of Cambridge physicist and theologian John Polkinghorne. Based on his interpretation of the Bible and his knowledge of physical cosmology and evolutionary biology, Polkinghorne is convinced that the eschatological resurrection of the dead cannot arise out of the natural world. Dead people stay dead. Unless, of course, God acts to raise them. Our hope rests solely in God. "An ultimate hope will have to rest in an ultimate reality, that is to say, in the eternal God himself, and not in his creation.[21]

The scope of the projected transformation is tantamount to the creation of the world in the first place. However, there is a difference. Whereas the present creation arrived *ex nihilo*, out of nothing; the new creation will be *ex vetere*, a transformation of the existing creation.

As we saw with Pannenberg earlier, key here is that resurrection of the individual is inextricably tied to the renewal of all things. "The new creation represents the transformation of that universe when it enters freely into a new and closer relationship with its Creator, so that it becomes a totally sacramental world, suffused with the divine presence." This suffusion of divine presence translates into healing. "Its process can be free from suffering, for it is conceivable that the divinely ordained laws of nature appropriate to a world making itself through its own evolving history should give way to a differently constituted form of 'matter', appropriate to a universe 'freely returned' from independence to an existence of integration with its Creator."[22] When we think of the healing miracles during Jesus' lifetime, we could think of his person "suffused with the divine presence." The healing power in Jesus' person pre-figures or pre-realizes the healing power of the entire new creation. "The *ultimate* destiny of creation, prefigured in the resurrection of Christ, [is] understood as the seed from which God's new creation has begun to grow through the redeeming transformation of the old creation, will indeed be a state in which God will be 'all in all' (1 Cor. 15:28), so I believe that panentheism will prove to be an eschatological reality."[23] The suffusion of the world with divinity at point omega implies, for Polkinghorne, eschatological panentheism.

---

[21] John Polkinghorne, 1993–1994 Gifford Lectures in *The Faith of a Physicist* (Princeton: Princeton University Press, 1994) or *Science and Christian Belief* (London: SPCK, 1994) 162–163.

[22] Ibid., 167.

[23] John Polkinghorne, *Faith, Science, and Understanding* (New Haven CT: Yale University Press, 2000) 90–91.

Even though Polkinghorne's hermeneutic explicates the faith within the horizon of modern science, he does not allow the scientific worldview to dictate exhaustively his explication. On theological grounds he argues that the present universe will have its material with its natural laws transformed. We will have new laws conducive to the avoidance of suffering and death. These new natural processes will be healing and life-giving because they will be suffused with the divine presence, the source of healing and life.

This leads to a distinction between evolution and eschatology. The new creation will not evolve on the basis of existing resources within the present creation, says Polkinghorne. Only an eschatological action on the part of God can make this transformation happen. The ground of hope "cannot lie in a mere evolutionary optimism, a belief that the unfolding of present process will lead to ultimate fulfillment, either individually or cosmically," writes Polkinghorne. "To put it bluntly, the only ground of true hope that there could be, within time and beyond it, lies in the everlasting faithfulness of the creator God."[24]

The corollary to this broad eschatology is that our immortality is not inherent in the human soul. Death is total annihilation of the psycho-somatic person, at least from our point of view. No quality within us as mortal beings has any purchase on immortality. If we are to rise from the dead, it will be due solely to the exertion of divine power. "The Christian hope is, therefore, for me not the hope of *survival* of death, the persistence *post mortem* of a spiritual component which possesses, or has been granted, an intrinsic immortality. Rather, the Christian hope is of death and *resurrection*."[25]

Is the Soul a Pattern of Information?

But, we might ask: what about contintuity? Will eschatological transformation create such newness that all connection with the past and present will be severed? Here is the way Polkinghorne handles it. He defines the soul as an information bearing pattern. He places himself in the tradition of Aristotle and Thomas, for whom the soul is the form of the body. What keeps Polkinghorne from garden variety Platonism is that it is God, not the soul on its own, that remembers our pattern. "I believe it is a perfectly coherent hope that the pattern that is me will be remembered by God and its instantiation will be recreated by him

[24] John Polkinghorne/Michael Welker, *Faith in the Living God* (Minneapolis: Fortress Press, 2001) 104.

[25] *Faith of a Physicist*, 163.

when he reconstitutes me in a new environment of his choosing. That will be his eschatological act of resurrection."[26]

On the one hand, death is complete, total. On the other hand, disembodied information perdures. The dissipation of our biological atoms into the surrounding physical environment is complete, total. Who we were physically in the past is not retrieved in its original substance. Resurrection consists of transformation into a spiritual body, the form of which is recalled by God into a new individual existence. God remembers us as information patterns.

I still would want to ask Polkinghorne why he limits continuity to the divinely remembered soul or pattern. I recall in Paul's vocabulary it is the body or *soma* that is repeated; no soul or *psyche* survives in the spiritual body, *soma pneumatikon* (1 Cor. 15:42–44). This text suggests that what will be changed is the material of the new creation making up the future resurrected body. On the one hand, Polkinghorne seems to offer an appropriate explication regarding materiality. The new material will obey new laws of nature, laws of life rather than laws of death. Yet, on the other hand, our identity is carried forward only by a divinely remembered information pattern, not physically. Be that as it may, such an idea of resurrection "is both immensely thrilling and deeply mysterious."[27]

As we have seen, Polkinghorne stresses the totality of death. This is because such totality is natural. Resurrection is not natural, at least as we understand nature today. Resurrection will require an act of God. "Thus the Christian hope centres on a real death followed by a real resurrection, brought about through the power and merciful faithfulness of God ... It is not necessary, however, that the 'matter' of these bodies should be the same matter as makes up the flesh of this present world. In fact, it is essential that it should not be. That is because the material bodies of this world are intrinsically subject to mortality and decay. If the resurrected life is to be a true fulfillment, and not just a repeat of an ultimately futile history, the bodies of that world-to-come must be different, for they will be everlastingly redeemed from mortality. Science knows only the matter of this world but it cannot forbid theology to believe that God is capable of bringing about something totally new."[28]

---

[26] Ibid.

[27] Ibid., 164. See: John Polkinghorne, "Eschatological Credibility: Emergent and Teleological Processes," in *Resurrection: Theological and Scientific Assessments*, ed. by Ted Peters/Robert John Russell/Michael Welker (Grand Rapids MI: Wm. B. Eerdmans, 2002) 43–55.

[28] John Polkinghorne, *Science and Theology: An Introduction* (London/Minneapolis: SPCK and Fortress, 1998) 115–116. One of the related issues not addressed in this chapter is the relationship of time to eternity. The Christian eschatological vision assumes a connection, because salvation and fulfillment consists of a transformation of

In summary, Polkinghorne thinks of the resurrected spiritual body as material, but not the same material we enjoy presently in our present biological make-up. We have a soul, defined as the form of the body, that is, the complex information-processing pattern of the body. This soul or pattern, however, is not in itself immortal. Rather, it supplies the form by which the divine mind remembers us and on the basis of which God in the new creation recreates us. Who we are in the new creation is who we are; but this continuity of identity is a gift from God.

This leads me to offer a slight modification of the Polkinghorne view. Polkinghorne places weight on the information pattern that constitutes the human soul and is remembered by God. I would prefer to place the weight on the *relationship* we temporal beings enjoy with our eternal God. Roland Chia follows this relationalist path with reference to the *imago dei* or image of God. "The image of God in humanity cannot be destroyed by sin or death. So long as God maintains his relationship with humanity, so long as he holds fast to humanity, the human person's designation as the image of God is indissoluble, inalienable, and immortal."[29] Mark Worthing reiterates this point well. "We are in the image of God because God relates to us. Not vice versa. The image of God must not only be understood in a primarily relational way, but also radically as gift."[30] The key, in my judgment, is God's relationship to us, within which the dynamism of transformation takes place.

The First Instantiation of a New Law of Nature

As we have seen, a historical retrieval of the event of Jesus' Easter resurrection leads immediately to spelling out its significance for the es-

---

the temporal world. Yet, as Paul Davies points out, it is difficult to conceive of a world going and going through more time after it has made its point, after its fulfillment. "I believe there is no plausible cosmological model on offer at this time that simultaneously satisfies two widespread but seemingly conflicting desires: that the universe has an ultimate destiny or purpose, and that the universe will exist interestingly for eternity. This fundamental tension between progress and eternity pervades the history of religion, and is solved in classical Christian theology ... by placing God ... outside of time altogether." Paul Davies, "Eternity: Who Needs It?" in *The Far Future Universe: Eschatology from a Cosmic Perspective*, ed. by George F.R. Ellis (Philadelphia/London: Templeton Foundation Press, 2002) 50.

[29] Roland Chia, "Biological Essentialism and the Person" in *Beyond Determinism and Reductionism: Genetic Science and the Person*, ed. by Mark L.Y. Chan/Roland Chia (Adelaide, Australia: Australian Theological Forum, 2003) 186–187.

[30] Mark William Worthing, "Human and Animal Intelligence: A Difference of Degree or Kind?" in *God, Life, Intelligence, and the Universe*, ed. by Terrence J. Kelly, S.J./Hilary D. Regan (Adelaide, Australia: Australian Theological Forum, 2002) 109.

chatological future of the created universe. If what happened in Jesus' resurrection is a prolepsis of what will happen to us upon resurrection into God's promised new creation, then we must speculate about a future world where dead people don't stay dead. What will resurrected life look like? Polkinghorne has made it clear that, according to present laws of physics, resurrected life is impossible. The laws of nature will need to be transformed.

This is the "problem of eschatology and cosmology," as Robert John Russell formulates it. This problem is "intensified to the breaking point if we take the resurrection of Jesus to entail 'bodily resurrection', since this view of the resurrection is coupled with an eschatology which entails the transformation of the present creation into a new 'heaven and earth'."[31] Russel assumes an inextricable connection between what happened to Jesus at Easter and what will happen to the laws of nature at the advent of the eschaton.

Now we need to ask Russell: did Jesus' resurrection abide by the current laws of nature, or the expected new ones? Russell answers with this: Jesus' Easter resurrection constitutes "the first instantiation of a new law of nature" (FINLON). The future new reality has arrived ahead of time, so to speak, in the singular event of Easter. What was true for Jesus on the first Easter will become true for all of physical reality at the advent of the eschaton.

Now is this just plain crazy? Are we talking about a transformation of physical reality so radical that it has no continuity with the present physical world? No. The physical world we currently live in has windows open toward such transformation, anticipations of newness. Specifically, the contingency inherent in natural events just might be a pointer toward eschatological newness.

FINLON belongs in the context of a larger discussion, the discussion of consonance vs. dissonance. When it comes to the concept of resurrection, we must admit that, instead of consonance, we see dissonance between science and theology. To put it crudely, any self-respecting scientist is likely to assume that it is a law of nature that dead people stay dead. Because every dead person we have seen remains dead, we would argue by analogy that Jesus must have remained dead – that is, there was no Easter.

So, we must admit dissonance between the Christian claim for resurrection and common scientific sense. Furthermore, the divinely promised new creation anticipated by theologians seems to be dissonant

---

[31] Robert John Russell, "Sin, Salvation, and Scientific Cosmology: Is Christian Eschatology Credible Today?" in *Sin and Salvation*, ed. by Duncan Reid/Mark Worthing (Adelaide, Australia: Australian Theological Forum, 2003) 131.

with the far future of the universe forecasted by physical cosmologists. Big Bang cosmology and correlative projections into the far future of the universe predict an eventual end to life and even an end to the material world as we know it. Scientists assume that the laws of nature presently in effect will determine what will happen to our universe a hundred billion years from now, perhaps forever. This could be interpreted as dissonant with Christian eschatology, because Bible reading theologians work with a divine promise that this world will undergo a transformation, that this creation will be renewed and redeemed. The scientific future appears irreconcilable with the theological future.

Russell provides two death scenarios, freeze or fry. Since the Big Bang 13.7 billion years ago, the universe has been steadily expanding. A key scientific question is this: is the universe open or closed? If it is open – that is, if the amount of mass is insufficient to stop the process, then it will continue to expand and continue to cool following the law of entropy. All the original heat will dissipate, and any remaining matter will fall into a state of equilibrium. In short, it will freeze out of existence. However, if the universe is closed – that is, if the amount of mass is above the relevant threshold, then at some point expansion will stop. Gravity will cause its motion to reverse, and all matter will reconverge on a central point, heating up on the way toward its doom in an unfathomably hot fireball. In short, it will fry. Whether freeze or fry, the future of the cosmos is finite. Whether freeze or fry, the scientific picture does not match the biblical picture of a new creation where the "wolf shall live with the lamb" (Isaiah 11:6) and where "death will be no more" (Revelation 21:4).

These points of dissonance between faith and science have led Russell to formulate FINLON. The context within which Russell first developed FINLON was an analysis of the concept of contingency, with special reference to the theory of the resurrection developed by Pannenberg.[32] Russell proposes a threefold typology of (1) global contingency; (2) local contingency; and (3) nomological contingency. Each of these has subtypes. The first, global contingency, yields (1a) global ontological contingency which asks: why does the universe exist *per se*? Why is there something and not nothing? Or, (1b) global existential contingency which asks: why does the universe have such global characteristics as the fundamental laws of physics and the constants of nature? Or, why this particular universe and not another one? The second, local contingency, yields (2a) local ontological contingency which asks: why does this par-

[32] Robert John Russell, "Contingency in Physics and Cosmology: A Critique of the Theology of Wolfhart Pannenberg," *Zygon*, 23:1 (March 1988) 23–44. See chapter one of the Festschrift for Robert John Russell, *God's Action in Nature's World,* ed. by Ted Peters/Nathan Hallanger (Aldershot UK: Ashgate, 2006).

ticular thing exist and continue to exist? Or, (2b) local existential contingency, which asks: why does this particular thing have the characteristics it has? When Russell turns to the third, nomological contingency, he locates the contingency of nature's laws. He asks: could the first instance of a new natural process inaugurate a new law of nature?

New laws of nature have already appeared in physical history. From the era of the Big Bang to the onset of life, no laws governing life were in effect. The onset of biological processes began only well into the history of the cosmos. Many aeons since the Big Bang 13.7 billion years ago had to pass lifelessly before the first sign of life on planet earth, only 3.8 billion years ago. With the arrival of life forms, so also came the arrival of life processes and the first instantiation of biological laws of nature. Nature has a history, and new chapters are genuinely new. A precedent has been set.

Now, does the resurrection of Jesus fit this paradigm? Might we think of what happened to Jesus on Easter as the first instantiation of a new natural process which we can expect to become universal at some point in the future? Within the category of nomological contingency Russell places the resurrection as a *first instantiation contingency.*

The first instantiation contingency is linked to our concept of prolepsis. This leads to an epistemological question: should one aspect of the eschatological future arrive ahead of time within our present eon still governed by what we acknowledge as nature's laws, could we discern it? If the eschatological new creation would include the resurrection of the dead, and if a person would rise from the dead now, could we discern it? Could we understand the Easter resurrection of Jesus as the first instantiation of a new law of nature that will apply universally at the advent of the renewed creation?

In order to discern it, says Russell, we need to suspend two impeding philosophical assumptions made by natural science, namely, analogy and nomological universality. According to the principle of analogy, the future will be just like the past. This means that, based upon analogy from observations about the natural world we make now we can project what the universe will be like thirty-five billion years from now. This needs critique. If in the near or distant future God acts to transform and redeem this world, it would be unpredictable according to the principle of analogy.

According to the principle of nomological universality, the same laws which govern the past and present will govern the future as well. This needs critique. Again, should God the creator re-create the world so that different laws obtain – such as wolves living with lambs and the elimination of death – then existing laws would no longer apply. The law that dead people stay dead would no longer apply. What we need to say about the Easter resurrection of Jesus, according to Russell, is that the future law that dead people will rise into the new creation has occurred ahead of time. Jesus is the "first fruits of those having fallen asleep" (1 Cor. 15:20). On Easter, God inaugurated a new law of nature, one that we will see become universal at some point in the future.

Russell finds distinctively theological warrant for holding this view. He claims "on *theological* grounds that the processes of nature that science describes are the result of God's ongoing action as Creator; their regularity is the result of God's faithfulness. But God is free to act in radically new ways, not only in human history but also in the ongoing history of the universe, God's creation."[33]

At the moment when God raised Jesus from the grave on the first Easter Sunday, the law that dead people remain dead was still in effect. It remains in effect today. In this context, this divine act leaves us with an anomaly. It could appear to be a miracle, a violation of existing laws of nature. However, Russell wants us to view that original resurrection as the first instance of a general resurrection. The advent of the new creation complete with general resurrection will establish that resurrection is a universal law. Jesus' Easter resurrection was the first instance, the prolepsis. Yet, it looks like a miracle.

The Life of Beatitude

The *Life of Beatitude* links God's ultimate future with the inner life or our individual soul and the outer life of communal living. Rudolph Bultmann stresses the existential power of this truth. "Eschatological existence has become a possibility for us through the fact that God has acted and made an end of the world as 'this world', by making us ourselves new. 'If anyone is in Christ, he is a new creation; the old has passed away; behold, the new has come'[34] (2 Cor. 5:17)." The future is present now in the faith of the believer.[35]

---

[33] Robert John Russell, "Bodily Resurrection, Eschatology, and Scientific Cosmology, in *Resurrection: Theological and Scientific Assessments*, 19.

[34] Rudolph Bultmann, *New Testament and Mythology and Other Basic Writings*, ed. and tr. by Schubert M. Ogden (Minneapolis: Fortress Press, 1989) 30. Bultmann poignantly describes the dynamism of faith, even though he would not endorse the objectivist element of proleptic eschatology I advocate. Neither would Gerhard Sauter, who wants to diminish the apocalyptic and utopian elements of eschatology in favor of enhancing the ability of faith to cope now, in the present moment, in a state of hope." Present and future cannot be distinguished as sharply as our grammar would have it: the future as the approaching not-yet, disappearing through the elusive moment of the now into the past, the no-longer. That the future of salvation enters into us means rather that we already have all we need to live with God." Gerhard Sauter, *Eschatological Rationality* (Grand Rapids MI: Baker Books, 1996) 26.

[35] The Life of Beatitude views time from the subjective experience of a person of faith with the future Christ present. What will be required of a theologian who con-

The presence now of future salvation lowers if not eliminates anxiety. Douglas John Hall emphasizes that our faith in the resurrection relieves us of the anxiety of facing our own death and liberates us for robust living. The Holy Spirit places within us the confidence to trust God, even in the face of doubts which might persist. "The fear of death, preoccupation with death, debilitating anxiety in the face of death, in short the sting of death – this, by grace, and *already*, is profoundly countered in this tradition. It is countered by *faith*, that is, by that trust in God wrested from us by the divine Spirit who struggles with our spirits, causing us despite our own doubt and fearfulness to say, 'Abba, Father' (Rom. 8:15) – that is, to consent."[36]

The future is also proleptically present in the Sacrament of the Altar. Even though our Eucharistic celebrations remember the night in which Jesus was betrayed, they also include the meal which anticipates the heavenly banquet Jesus spoke of in his parables of the kingdom. What we mean by the "real presence" of Christ includes eschatological presence. "Even if Christ's future is not yet completed," writes Michael Welker, "we have in his presence salvation in its entirety and complete comfort and encouragement."[37] Liturgically we sing of the bread and wine as a "foretaste of the feast to come."

Jürgen Moltmann believes this faith produces a life of hope. "Christian faith is faith in the resurrection, and the resurrection is literally just that: rising up again. It gives us the strength to get up, and the creative freedom to begin something once more in the midst of our ongoing history, something fresh. *Incipit vita nova* – a new life begins. That is the truly revolutionary power of hope."[38] Welker puts it this way: "The advent of the reign of God can indeed be experienced in the spread of love, mercy, forgiveness, and true justice."[39]

We believers live in the era of the church. We live like an arc between two terminals, yesterday's Easter resurrection and tomorrow's resur-

---

structs a worldview based upon prolepsis would be the development of an appropriate concept of time that would include retroactive ontology. The construction of a concept of time is difficult, if not impossible, according to Antje Jackelèn, because we inescapably view time's passage from our situation in life. We dare not construct a simple linear view of time; but rather incorporate dynamism and relationality. Yet, it is not yet clear how to proceed. *Zeit und Ewigkeit: Die Frage der Zeit in Kirche, Naturwissenschaft, und Theologie* (Neukirchen-Vluyn: Neukirchener, 2002) 306.

[36] Douglas John Hall, *The Cross in Our Context* (Minneapolis: Fortress Press, 2003) 214.

[37] Michael Welker, *What Happens in Holy Communion?*, tr. by John Hoffmeyer (Cambridge UK/Grand Rapids MI: SPCK and Eerdmans, 2000) 119.

[38] Jürgen Moltmann, *In the End – The Beginning: The Life of Hope* (Minneapolis: Fortress Press, 2004) xi

[39] Welker, *Holy Communion,* 119.

rection into the eschatological new creation. What happened to Jesus on the first Easter was a prolepsis of what will happen to the entirety of the created universe at omega. God's promise of a new creation will necessarily include changes in the laws of nature to accommodate resurrected existence and harmony among all life forms. What we see in the Easter resurrection is Jesus' incarnation of the first instantiation of that eschatological law of nature where the dead will rise and come to full flower in the kingdom of God. To live now in anticipation of our future resurrection is to live the life of Beatitude.

Chapter 3

Barriers to Grace in a Scientific Era

Constructive theology for heirs of the Protestant Reformation, especially the Lutheran branch, is built on a foundation of grace. Although the Reformation fought battles over grace and works, warriors during the Enlightenment skirmished over a different set of issues. Modern science, the child of the Englightenment, simply does not address the question of grace directly. Yet, it cannot be avoided.

Theologians were conscripted into the army of grace during the sixteenth century Reformation. They needed to defend grace against merit, works righteousness, faith formed by love, and Tridentine statements such as, "If anyone shall say that by faith alone the sinner is justified, so as to understand that nothing else is required to cooperate in the attainment of the grace of justification, and that it is in no way necessary that he be prepared and disposed by the action of his own will: let him be anathema."[1]

Christian theology needed a reconstruction, the pouring of a new foundation in bedrock. By the end of the twentieth century, what was being built was much more sturdy. "Together we confess," said Lutheran and Roman Catholic theologians together, "by grace alone, in faith in Christ's saving work, and not because of any merit on our part, we are accepted by God and receive the Holy spirit, who renews our hearts while equipping and calling us to do good works."[2]

It appears that the underpinnings of grace have been reestablished. We are now free to rebuild the house of systematic theology on a firm foundation. What this means is that explication of remaining theological *loci* should cohere as complementary articulations of the graciousness of the God in whom we put our faith. Everything we say should be grace imbued.

This will not be easy. Near invisible barriers obstruct easy construction. Of course, we are familiar with perennial barriers in anthropolo-

---

[1] The Council of Trent, Decree on Justification, can 9. See: *The Condemnations of the Reformation Era*, ed. by Karl Lehmann/Wolfhart Pannenberg (Minneapolis: Fortress, 1990) 34.

[2] *Joint Declaration on the Doctrine of Justification, par. 15*, The Lutheran World Federation and the Roman Catholic Church (Grand Rapids: Eerdmans, 2000) 15.

gy such as the human propensities for self-justification and hypocrisy. Here I would like to look at two others, two other impediments if not barriers to an easy construction of a grace-based theology: doubt and evil. The form that reflection on doubt and reflection on evil take in contemporary culture do not appear automatically on the theological agenda as questions for which grace is the answer. The form in which these theological questions are ordinarily posed presuppose an understanding of God that disregards the question as to whether God is gracious or not. If doubt and evil are to be given room in theological construction as grace oriented theologians envision it, we need reformulations so as to make the foundation in divine grace visible.

In what follows, my point will be that the concept of grace refers us first and foremost to a quality of God, namely, that God is loving, and this love is unilateral and unconditional. Exactly what this means for contemporary theological thinking and for cultural thinking remains to be explicated. Once this point is made, I would like to turn to two barriers to a healthy understanding of grace: doubt and evil. On the one hand, both doubt and evil are perennial opponents to God's gracious activity. On the other hand, contemporary intellectual life provides them with recognizable uniforms on the cultural battlefield. Doubt wears the uniform of a scientifically based agnosticism and even atheism; and evil wears the uniform of philosophy of religion as it raises the theodicy problem. The theologian of grace should make ready to jump the barriers.[3]

Sola Gratia, Sola Fide

By 'grace' we refer to the "unmerited assistance from God, especially as it pertains to salvation," in the words of David Yeago.[4] The final phrase here, "as it pertains to salvation," recognizes the original context in which Reformation Lutherans received their charism. As we shift from the sixteenth to the twenty-first century, the terrain has shifted somewhat away from salvation and back to creation. Not that salvation has been forgotten, to be sure; yet the challenge is to extend our boot camp preparation in salvation to *loci* such as God and creation.

As it pertains to salvation, Martin Luther found himself analyzing the dynamic interaction between grace and faith like a Swiss watchmaker

---

[3] This chapter is a revision of my article "Grace, Doubt, and Evil: The Constructive Task of Reformation Theology," *Dialog*, 41:4 (Winter 2002) 273–284.

[4] David S. Yeago, "Grace," *The Oxford Encyclopedia of the Reformation*, ed. by Hans J. Hillerbrand (4 Volumes: New York/Oxford: Oxford University Press, 1996) 2:184.

analyzes each and every gear in a an elegant time piece. Among other things, the spring for faith is an "alien righteousness." Faith receives justification and salvation from beyond, from God. "This alien righteousness, instilled in us without our works by grace alone – while the Father, to be sure, forwardly draws us to Christ – is set opposite original sin, likewise alien, which we acquire without our works by birth alone. Christ daily drives out the old Adam more and more in accordance with the extent to which faith and knowledge of Christ grow."[5]

What makes this alien righteousness effect salvation in us is the presence of Christ placed in our faith by the Holy Spirit. By God's grace, Christ is present in our faith. His presence is that alien righteousness. True faith "takes hold of Christ in such a way that Christ is the object of faith, or rather not the object but, so to speak, the One who is present in the faith itself."[6] Faith justifies and hence saves, according to Luther's theology, because Christ is himself present in our faith. Proleptically speaking, our future salvation is present within us now, ahead of time.

The presence of Christ and the accompanying alien righteousness effect salvation through what has come to be known as the happy exchange (*fröhlicher Wechsel*). "Christ is full of grace, life, and salvation. The soul

---

[5] Martin Luther, "Two Kinds of Righteousness" in *Luther's Works*, American Edition, Volumes 1–30 ed. by Jaroslav Pelikan (St. Louis: Concordia, 1955–1967), Volumes 31–55 ed. by Helmut T. Lehman (Minneapolis: Fortress, 1955–1986) 31:299. Here in 1519 Luther saw the dynamic of grace's effect on faith in terms of growth or even progress. "For alien righteousness is not instilled all at once, but it begins, makes progress, and is finally perfected at the end through death." Ibid. As Gerhard Forde develops the theology of justification by faith, he decries the notion of growth or progress. Grace at the beginning must be final, he emphasizes. "In its simplest form the problem may be stated thus: If justification conceived as forgiveness comes at the beginning of the process, the process is superfluous ... if, on the other hand, justification comes at the end of the process, justification is superfluous ... Both the divine act of justification and the human process of becoming just according to the law cannot simultaneously be real." Gerhard Forde, "The Christian Life," in *Christian Dogmatics*, ed. by Carl E. Braaten/Robert W. Jenson (2 Volumes: Minneapolis: Fortress, 1984) 2:404. The instillation of salvific alien righteousness, according to Forde then, is immediate and total; righteousness or justification is not distributed bit by bit over time. Despite what Luther says, is this what Luther actually means?

[6] Luther, "Lectures on Galatians" of 1535, *Luther's Works*, 26:129. Tuomo Mannermaa recognizes that the role of the indwelling Christ in faith is central to Luther but not necessarily to Lutherans. "According to Luther, Christ (in both his person and his work) is present in faith and is through this presence identical with the righteousness of faith. Thus, the notion that Christ is present in the Christian occupies a much more central place in the theology of Luther than in the Lutheranism subsequent to him. The idea of a divine life in Christ who is really present in faith lies at the very center of the theology of the Reformer." "Why is Luther So Fascinating? Modern Finnish Luther Research," *Union With Christ*, ed. by Carl E. Braaten/Robert W. Jenson (Grand Rapids: Eerdmans, 1998) 2.

is full of sin, death, and damnation. Now let faith come between them and sins, death, and damnation will be Christ's while grace, life, and salvation will be the soul's."[7] We are deemed just and hence justified, because Christ is just and we have received his justness through this exchange of properties.

At stake here is something more important than death. It supersedes death in ultimacy. In fact, justification by faith incorporates death and life, death to a sinful state of existence and resurrection to eternal life with God. The promise of resurrection is the future hope of the Christian gospel; yet, there is more. That future resurrection can be present in faith now because the Easter Christ is present in faith. The gospel means "death and new life in the crucified and risen Christ," writes Gerhard Forde.[8] The Christ present to faith has already died, so his death becomes our death. The Christ present to faith has already risen from the dead, so his resurrection becomes our resurrection. This makes the pronouncement of grace unconditional. "To the age old question, 'What shall I do to be saved?' the confessional answer is shocking: 'Nothing! Just be still, shut up and listen for once in your life to what God the Almighty, creator and redeemer, is saying to his world and to you in the death and resurrection of his Son! Listen and believe!'"[9] Because God has acted toward us in an unconditioned way, we speak of salvation as *sola gratia*, by grace alone. Because faith here refers less to a human activity and more to the dynamic of Christ's presence in the life of the believer, we speak of salvation as *sola fide* and *solus Christus*, by faith alone and by Christ alone.[10]

This presence of the resurrected Christ empowers the Life of Beatitude. Omega is present in today's faith.

Grace as God's Favor Towards the World

Worthy of note here is the locus of the activities of grace. We tend to presume that grace is important "as it pertains to salvation." The location of

---

[7] Luther, "The Freedom of a Christian," *Luther's Works*, 31:351.

[8] Gerhard O. Forde, *Justification By Faith – A Matter of Death and Life* (Minneapolis: Fortress, 1982) 3.

[9] Ibid., 22.

[10] If the phrase 'justification by faith' implies salvation achieved as the result of human faith as a human work, complains Paul Tillich, then we risk a misleading interpretation. "The cause is God alone (by grace), but the faith that one is accepted is the channel through which grace is mediated to man (through faith)." *Systematic Theology* (3 Volumes: Chicago: University of Chicago Press, 1951–1963) 2:279. He suggests using the phrase 'justification by grace through faith' to emphasize that salvation is a divine, not a human, work. When properly understood as involving the presence of Christ, the biblical phrase 'justification by faith' sufficiently implies *sola gratia*.

grace is its action in the human soul and in the life of the church. The interaction of grace and faith takes place in human subjectivity, in the faith of the individual and in the church, the *communio sanctorum*. Lutherans, Anglicans, and Roman Catholics affirm an objective component, namely, the real presence of the body and blood of Christ in the elements of the eucharist, to be sure; yet the arena of action is the subjective domain within persons of faith. Does this exhaust the scope of grace?

What about other domains, such as the attributes of God or the created order of nature? Certainly we consider the loving disposition within the divine life to be the fountain of all God's gracious activity; and the physical world God has created is a sign of God's glory and a means of divine grace. Grace is God's favor, and God favors the world.

Grace is a quality that belongs first and foremost to God. "I take grace in the proper sense of the favor of God," remarks Luther, contrasting it with a quality of the human soul.[11] Joseph Sittler reiterates that grace is rooted in God's love, and it is expressed toward creation: "The fundamental meaning of grace is the goodness and loving-kindness of God and the activity of this goodness in and toward his creation."[12] What we know as the temporal creation is a gracious act deriving from God's eternal love. "Love is the only real answer we have to the startling question, Why should there be anything at all rather than nothing? Love grants existence and grants it contingently," writes Wolfhart Pannenberg.[13]

Thinking of grace in terms of God and creation expands our frame of reference. We, today's people of faith, are but a brief paragraph within the chapter on the church, which in turn is but an episode in the epic narrative of God's creation stretching back to a Big Bang beginning and forward to a transformative new creation. The Christian gospel is the message that all of this is graced. "The gospel reaches backward and forward all along the line from creation to consummation because Christ is the eschatological revelation of God already at the beginning of things," writes Carl Braaten; "The world was created through Christ, and all things will ultimately reach their end in him as the Judge and Lord. This is the biblical meaning of calling Christ the *alpha* and the *omega*."[14]

The divine grace of which we speak, in summary, has certain characteristics. First, the word 'grace' is a descriptor for a divine disposition,

---

[11] Luther, *Luther's Works*, 32:227.

[12] Joseph Sittler, *Essays on Nature and Grace* (Minneapolis: Fortress, 1972) 24.

[13] Wolfhart Pannenberg, *Theology and the Kingdom of God* (Louisville: Westminster John Knox Press, 1969) 65.

[14] Carl E. Braaten, *Principles of Lutheran Theology* (Minneapolis: Fortress, 1983) 111.

namely, the disposition for love that leads to both creation and redemption. Second, grace is unconditional; it originates solely in divine *aseity* and at God's initiative. Third, in our faith grace is experienced as a gift, not as a reward for works accomplished or merits earned. Fourth, the content of this gift is the presence of Jesus Christ, the uninvited but very welcome guest from heaven dwelling with us on earth.

The sixteenth century Reformation was rocked by controversies over the role of grace in justification and salvation; and Lutheran theology was born with a special vocation to champion the cause of *sola gratia* within Christendom. The former controversies have not entirely disappeared, to be sure; yet both the church and the world of the twenty-first century provide a significant shift in context. If to be a friend of grace is the distinctive Lutheran charism, then this new context extends the call to penetrate more deeply into understanding that grace by discerning spirits – by discerning the supports and barriers to grace – in the *Geist* of the new millennium.

Barriers to Grace

It is a most curious exercise to entertain the idea that barriers would stand in the way of grace, that grace would fail to prompt a fully paved road to welcome it. It seems so unlikely. If grace refers to a gift freely given, how could this possibly elicit a negative response? Gifts are almost universally welcome; and gratitude is the normal response. Yet, the drama of biblical history and subsequent history demonstrates that divine love stirs up human anxiety, divine giving stirs up human pride, divine care stirs up human self-assertion, and divine presence leads to crucifixion and death. One of the challenges to God in this drama is for divine grace to turn human enemies into friends.

This biblical drama reminds us that existing friends of grace have two overlapping tasks. First, theologically, we need to penetrate surface images to an underlying understanding of human nature. We need to assess theologically just why we human beings have a predisposition to resist grace, a propensity to repudiate the God who unconditionally loves us. Second, we need to survey the contemporary cultural landscape to identify the topography of intellectual, spiritual, and political landmarks where grace has no safe haven. We need to mount a theological apologetic, speaking directly to contemporary discontent within the church and within the world. This chapter will deal primarily with the latter task.

Yet, before we turn to our contemporary context, a word about human nature is in order. In the New Testament the word for 'sin' (*hamar-*

tia) connotes missing the mark, as an archer may miss the targeted bull's eye. What is positive about this metaphorical image is that we human beings are aiming at something, perhaps aiming at a target God has placed before our eyes. Note in the quotation from Luther above that original sin is a form of alien unrighteousness just as what grace imparts is an alien righteousness. This could imply that our deepest human inclination is to aim for what is holy and divine; but our trajectory misses the mark and strikes a substitute for the true God. With Augustine's helpful insights, we understand this substitute for the true God to be ourselves. We are, said the Bishop of Hippo, turned in upon ourselves; we are *homo incurvatus in se*. By missing God, we hit ourselves. God's grace can straighten out the arrow so that the bull's eye gets hit, according to Augustine.[15]

If we draw anthropological corollaries from the controversies over justification by faith, we can posit that we human beings have a built in propensity for self-justification. The medieval derailing of Christian spirituality by the merit system should not have been unexpected; the pressure will always be with us to build our own ladder to heaven, our own Tower of Babel, and then climb up. Deep within the human soul is a sense of justice; actually, an inchoate sense that justice represents what is eternal. The imaginary ladders we build are steps toward justice; we then climb up and declare ourselves just. As temporal beings pursuing eternal life, consciously or unconsciously, we seek to appropriate justice to ourselves; because we presume that in justice is power, and in power is life. *Self-justification* becomes the term for describing the fruit of our passion for ourselves, for establishing our security and our stature in our public image and before God.[16]

Here, then, is the unavoidable conflict. If justification is given us by God, then self-justification is unnecessary. If justification is a gift, then our earning it is unnecessary. If justification is the result of an alien righteousness, then possessing it as an attribute of our own is unnecessary. If justice is something we wish to own for ourselves rather than borrow from Christ, then we have located a point of conflict. We must conclude that a potential for resistance to God's grace inheres in universal human nature; so we should not be surprised to find such resistance occasionally actualized in human spirituality, religion, and culture.

---

[15] "The fault in man is contrary to nature, and is just that which grace heals ... not that nature is a denial of grace, but that grace is the mending of nature." Augustine, *The Spirit and the Letter*, 47 (xxvii).

[16] Such an analysis of self-justification is taken up in chapter 6 of *Sin: Radical Evil in Soul and Society*, by Ted Peters (Grand Rapids: Wm. B. Eerdmans, 1994).

With this in mind, I would like to turn to two barriers resisting grace: doubt and evil.[17] Like a natural geyser, resistance to grace lies under pressure well beneath earth's surface and then gushes forth when an opening appears. Resistance to grace can spurt up without warning within the privacy of our individual psyches, within the spiritual customs of our religion, or within secular realms of self-understanding.

Doubt in a Scientific Era

Doubt in our era is sometimes formulated as a question, "does God exist?" Those who find it difficult to affirm a positive answer may attach a bumper sticker to their car, "Dog is my co-pilot." ("Dog" is "God" spelled backwards.) Beyond bumper stickers we find a range of tacit assumptions and articulate ideologies about the non-existence of God voiced in various forms of naturalism, materialism, scientific humanism, and overt atheism. Short of atheism, most prevalent in modern intellectual culture is agnosticism, the position of considered doubt. The term *agnosticism* comes from Thomas Huxley in 1869 and reflects the image modern science has of itself, namely, as modern it replaces the religion of the past. Using the metaphor of maturity, Huxley says: "When I reached intellectual maturity and began to ask myself whether I was an atheist, a theist, or a pantheist … I found that the more I learned and reflected, the less ready was the answer. They [believers] were quite sure they had attained a certain 'gnosis,' – had, more or less successfully, solved the problem of existence; while I was quite sure I had not, and had a pretty strong conviction that the problem was insoluble."[18] Maturity, it is assumed here, refers to an open-minded agnosticism that has outgrown the narrow-minded dogmatism of premodern religion.

Agnosticism both depends on yet fosters naturalism. According to modern naturalism, nature is the only reality there is. The natural realm is self-sustaining, self-regulatory, and even self-explanatory to those who know how to understand it, namely the scientists among us. The world of nature has no windows that open toward another world; no such thing as transcendence can be relied upon to ferry us beyond the material shore of existence. We simply cannot know any-

[17] I first introduced discussion of these four as "hurdles" to be jumped in an editorial, "Barriers to God's Grace," *Dialog*, 37:2 (Spring 1998) 82–83. In this editorial I listed four enemies: doubt, evil, legalism, and hypocrisy. Legalism and hypocrisy are still with us, and they also cry out for an updated analysis.

[18] Cited by Michael Shermer, *How We Believe* (New York: W.H. Freeman, 2000) 7.

thing about a noumenal reality that allegedly transcends the phenomenal world which we can study through science.

Even though religious people among us speak of heavenly realities, skeptics doubt that such speech refers to anything heavenly. What appears clear since Feuerbach and Freud, Marx and Lenin, is that religious symbols express human wishes, that religious visions are unconscious projections onto heaven of wishes for fulfillment that are unattainable on earth. What exists is human subjective desire, to be sure; what is doubted is the divine object of that desire. What needs explanation, contend the skeptics, is not God but rather religion.

Evolutionary theory has in recent decades become the materialist explanation of choice. Fields such as sociobiology and evolutionary psychology are trying to develop a comprehensive naturalism that will explain theological claims about God exhaustively in terms of evolutionary principles such as adaptive advantage. Religion, some claim, increases procreative advantage for some human groups over others; therefore, the religions that have survived to the present day must be the most reproductively successful. According to this theory, religion provides ethics as a cultural justification and inspiration that enhances our biological drive to produce children, an epigenetic tool in the construction of a preferred gene pool. Robert Wright speaks for the field of sociobiology here: "The reason they want children is because their genes 'want' children ... why fight it?"[19] Because of this, says Harvard entomologist E.O. Wilson, "Ethical and religious beliefs are created from the bottom up, from people to their culture. They do not come from the top down, from God or other nonmaterial source to the people by way of culture."[20]

What we find here is an alternative explanation for religious beliefs about God and about human values. It is a purported *scientific* explanation. The scientist says that we should abandon asking the theologian to explain religion; rather, we should ask the scientist to provide a more accurate understanding of why people believe in God or adhere to ethical principles. The underlying agnosticism with regard to matters divine justifies drawing a strictly mundane or this-worldly picture of religious believing.

---

[19] Robert Wright, *The Moral Animal* (New York: Pantheon, 1994) 148.

[20] Edward O. Wilson, *Consilience: The Unity of Knowledge* (New York: Alfred A. Knopf, 1998) 247.

From Agnosticism to Pluralism

Agnosticism supports religious pluralism. The impetus derives primarily from the opposition to perceived dogmatism characteristic of traditional religion, an opposition thought to be supportive of open-minded science. "I believe that gods exist to the extent that people believe in them. I believe that we created gods, not the other way around," writes the director of the Skeptics Society, Michael Shermer; "whether God *really* exists or not is, on one level, not as important as the diverse answers offered from the thousands of religions and billions of people around the world ... My only gripe with religion is when it becomes intolerant of other peoples' beliefs ... or the cultural suppression of diversity."[21] The impetus for religious pluralism among such skeptics and agnostics does not derive from a desire to pursue theological knowledge. They are not traveling many roads up a single mountain – that is, no motive exists here to study a large number of religious claims in order to accumulate increased knowledge of transcendent reality. Rather, the positing of pluralism has a political purpose, namely, to disestablish a dominant religion. Pluralism protects the right of each religious tradition to create its own projection of gods. And the agnostic secretly knows that all such religious projections are nothing but imaginary fictions.

In some instances, agnosticism can make way for full fledged atheism. Atheism is the positive assertion that no transcendent reality answers to the name of God. Paul Kurtz, avowed secular humanist and editor of *Free Inquiry,* makes this point by asserting that no God is listening to our prayers and supplications. "Prayers to an absent deity ... merely express one's longings. They are private or communal soliloquies. There is no one hearing our prayers who can help us. Expressions of religious piety thus are catharses of the soul, confessing one's fears and symbolizing one's hopes. They are one-sided transactions. There is no one on the other side to hear our pleas and supplications."[22] There is no grace for atheism, because there is no God who could be gracious

[21] Shermer, *How We Believe*, 11, xiii.

[22] Paul Kurtz, *Transcendental Temptation* (Buffalo: Prometheus Books, 1988) 22. "The presupposition of contemporary Western atheism lies in the development of modern natural science and its mechanistic picture of the world ..."; Wolfhart Pannenberg, "Types of Atheism and their Theological Significance," *Basic Questions in Theology* (2 Volumes: Minneapolis: Fortress, 1970–1971) 2:184. See also: Matt Donnelly, "The Twilight of Atheism: Examining a Once-Triumphant Worldview," *Science and Theology News,* 6:4 (December 2005) 12–14.

What appears on the surface here is that the question looks like this: "does God exist?" The emergence of the scientific method with its antipathy towards religious dogmatism and the emergence of a naturalistic worldview with windows closed to the transcendent make this a question of wide cultural significance. Yet, would it make sense to ask another question: "is God gracious?" It would matter little if we could find a way through empirical science or philosophical reason to demonstrate beyond a doubt that God exists in the sense that nature would have a creator and sustainer. As dramatic as such a proof might be, merely establishing a designer for nature falls short of establishing what is most vital to us existentially. In a curious sort of way, the question of the existence or non-existence of God takes second place to the question of God's graciousness. It is relatively unimportant to us to know whether or not a celestial engineer designed the natural world. What makes the difference in our lives and in our destinies is the presence or absence of divine grace.

Doubt Within Faith

Paul Tillich positively incorporates doubt into his concept of faith. An honest agnosticism, thinks Tillich, derives from a commitment to truth. When evidence is insufficient to be convincing, a person committed to truth must withhold judgment. Because knowledge of God differs radically from knowledge of things of this world, convincing evidence comes in insufficient supply to provide apodictic knowledge of God. Some measure of doubt is unavoidable, even healthy, especially when it is serious doubt accompanied by a sense of ultimacy. It is healthy because it exhibits faith in the truth, and true faith is trust in the God of truth. Indirectly, doubt is a form of faith; it is trust in the same truth that faith knows as God. "Serious doubt is confirmation of faith."[23]

Yet, this is insufficient as an apologetic because it deals only with the subjective side of the ledger, with a person's disposition toward pursuing what might or might not be true. What remains to be addressed is the objective question, "if there is a God, is this God gracious?" One resource Lutheran theologians can bring to the question is the Theology of the Cross (*theologia crucis*). "The manifest and visible things of God are placed in opposition to the invisible, namely, his human nature, weakness, foolishness ... it does him no good to recognize God in his glory and majesty, unless he recognizes him in the humility and shame of the cross ... God can be found only in suffering and

[23] Paul Tillich, *Dynamics of Faith* (New York: Harper, 1957) 22.

the cross," says Luther.[24] According to this Theology of the Cross, the glory and majesty of the invisible creator is hidden behind the visible phenomena of suffering and death. In addition, this suffering and death belong to the reality of God proper; they are manifestations of the invisible God's gracious presence within a visible world.

Among the many things the Theology of the Cross tells us is this: pursuing knowledge of God by speculating on the origins and design of the natural world is a diversion, whether successful or frustrating. The best it could yield would be knowledge – more than likely projected speculation accompanied by doubt – about an intelligent designer or celestial engineer. Only by looking at the cross – the finite, historical, this-worldly cross – does God's grace become visible. Only in the cross can we see God present as the one who shares our suffering, who becomes the victim of our sin, who by this very acceptance of victimage bestows forgiveness and by promising resurrection bestows eternal life. Only in the cross can we see an eternal love coming to temporal expression.

We may be dealing with three types of hiddenness here. The first is God's hiddenness from what scientists can see in nature. This is due to a methodological commitment on the part of the scientific community to limit their explanations to natural explanations. Using theological vocabulary, scientists study only the domain of secondary causation, not primary. Actually, this form of divine hiddenness goes back to a philosophical concern for apprehending an infinite God from a finite perspective, a noumenal reality from the world of phenomena. This form of hiddenness is the basis for modern agnosticism and atheism.

The second is the hiddenness of God's grace, God's care for us in our human plight. This is where Luther felt the tyranny of *deus absconditus*, the God of majesty who foreordains our creation and our destiny in sheer mystery. Is this high God indifferent to our concerns? Why doesn't this God answer prayer the way we ask?

Then, as Lois Malcolm argues, there is a third kind of hiddenness: the hiddenness of Christ in faith. Malcolm turns to an "apocalyptic epistemology" to refer to the light shed by a promised future transformation, a future reversal of darkness and light. God now "sees" in the depth of creation what we can only hope for by faith, a new reality. "The full redemption we await lies in the future. Evil and suffering have not ceased … Christians believe that the very redemption of the world (the primary sense of hiddenness) takes place in *creaturely* flesh. This, then, is the challenge of hiddenness language: to confess to the

[24] Luther, *Luther's Works*, 31:52–53.

goodness of reality and God's redemptive purposes for it even when it is hidden by the brokenness and horrors of life."[25]

Looking at the cross will not provide the empirical evidence or philosophical proof sought by agnostics of the modern scientific era. But, then, the cross does not purport to deliver what agnosticism is searching for. Rather than prove the existence of God metaphysically, the cross provides an oblique peek into the realm of transcendence and then turns our attention around so as to see the presence of the transcendent God in the most mundane of the mundane, the physical world replete with finitude, suffering, and death.

Evil and the Challenge to Divine Love

Suffering and death bring us to the problem of evil. According to philosophers of religion, the problem of evil combines doubt about God's existence with doubt about God's grace. Skeptical philosopher David Hume lays down the gauntlet: "Is he [God] willing to prevent evil, but not able? Then is he impotent. Is he able, but not willing? Then is he malevolent. Is he is both able and willing? Whence then is evil?"[26] Raised as a problem of logic, philosophers of religion frequently point out the incompatibility of three propositions:

God is omnipotent (all powerful).
God is omnibenevolent (all loving).
Evil and suffering exist.

As Hume points out, the affirmation of any two of these negates the third. Contemporary atheistic philosophers find the irresolvability of the *problem of evil* to be support for the denial of God's existence. Theologians who wish to defend the existence and the righteousness of God, find themselves in the position of proffering a *theodicy*, an argument that justifies God in the face of the presence of evil and suffering.

What the logical problem alludes to without capturing the depth of existential concern is the anguish we undergo when faced with the ugliness of evil and the horror of suffering. Whether natural evil in the form of suffering from disease or disaster, or historical evil in the form of human violence, we find ourselves victims of suffering we did not invite nor wish to endure. Agonizing questions are flung heaven-

---

[25] Lois Malcolm, "A Hidden God Revisited: Desecularization, The Depths, and God's Sort of Seeing," *Dialog*, 40:3 (Fall 2001) 189.

[26] David Hume, *Dialogues Concerning Natural Religion*, ed. by Richard H. Popkin (Indianapolis/Cambridge: Hackett, 1980) 63.

ward. Why do so many children who desperately need caring families and loving nurture find themselves trapped in cycles of abuse leading to torture and scarring for life? Why would European civilization with all its industrial advances and democratic achievements fall into wanton self-immolation by mustard gas and saturation bombing over two world wars? Why would a nation with a millennium of Christian influence devise a secret police force, death camps, and gas chambers? Why would the Japanese army massacre a quarter million Chinese in Nanjing, or al-Qaeda incinerate three thousand of our globe's citizens in New York's World Trade Center? Why would the twentieth century chronicle genocide after genocide: the Ottoman Turk genocide of Armenians; Nazi eradication of people with disabilities, Jews, gypsies, and enemies of the state; Bosnian Serb massacres of non-Serbs, Hutu extermination of Tutsis?

Genocides are the result of human sin; we refer to them as *human evil*. But, in our own era, *natural evil* also challenges religious commitments to a loving God. The controversies over evolutionary biology, beginning already with Charles Darwin himself, include the observation that, given the overwhelming amount of suffering by sentient beings in the predator-prey struggle and the extinction suffered by the vast majority of species due to the cold law of natural selection, no God could be considered responsible for creating such a brutal and bloody world of nature. And, if God is in fact responsible for creating the world ruled by natural selection, such a God would be morally unworthy. Niels Henrik Gregersen poses the challenge: "In two ways has the problem of theodicy been sharpened since Darwin, first because the existence of pain and suffering can no longer be explained by human sin, and secondly because the existence of what we (as ethical primates) cannot but term as brutal, can no longer be seen as accidental to evolution. The struggle for life is pre-human in origin and is built into the very way in which the world is wired for complexification. Wired by God?"[27]

Marilyn McCord Adams would refer to such genocides and such suffering in nature as "horrendous evils." With this term she refers to victimage so destructive that, it would seem, no future redemption could reinvest the victim's life with meaning. "*Horrendous evils* … I define … as evils the participation in which … constitutes prima facie reason to doubt whether the participant's life could … be a great good to him/her on the whole."[28] In the middle of anguish over evil, it ap-

[27] Niels Henrik Gregersen, "The Cross of Christ in an Evolutionary World," *Dialog*, 40:3 (Fall 2001) 200.

[28] Marilyn McCord Adams, *Horrendous Evils and the Goodness of God* (Ithaca/ London: Cornell, 1999) 26.

pears that the depth of suffering is so overwhelming that no incorporation into a more meaningful whole could provide healing.

Evil and the Free Will Defense of God

Some theologians have dared to try to address the existential anguish by resolving the logical conundrum. One solution is to deny being to evil, to render evil something that has no being in itself. Augustine tried taking this route. Evil is the privation of what is good (*privatio boni*) he said.[29] It may work philosophically to say that evil is a parasite off the good and deny to it any independent being; but the way we experience evil is as a structure of destruction, as the force of nonbeing destroying what is.[30] Whether it has being or not, evil is real and we reel from the suffering it causes. (As indicated in chapter one, the evil we experience as the force of destruction is nonbeing in its *me on* rather than *ouk on* form.)

Another of Augustine's tactics has become known as the free will defense. In principle Augustine affirms all three propositions, but by adding free will he transmutes evil into an expression of divine omnibenevolence. The creation of free creatures is a higher achievement than the creation of mere things such as robots or automatons, the argument goes; but when God bestows freedom upon us the risk arises that we might engage in evil and cause suffering. This is just what happened historically, says Augustine. Our primordial parents, Adam and Eve, voluntarily (*voluntas*) chose to turn away from God; they turned away from the eternal and immutable goodness of God toward the temporal and mutable goods of this world. From then on their choices and hence our choices (*arbitrium*) are made in behalf of ourselves, not God, because we are now *homo incurvatus in se*. We began with a free will and lost it to sin; later, God will restore us at a higher level of freedom where we will be freed from the inclination to sin. "God's foreknowledge had anticipated both – that is to say, both how evil the man whom he had created good should become, and what good he himself should even thus derive from him ... evils are so thoroughly overcome by good that, though they are permitted to exist for the sake of demonstrating how the most righteous foresight of God can make a good use even of them ... And evil is removed, not by removing any nature, or part of a nature, which had been introduced by the evil, but by healing and correcting that which had been vitiated and depraved.

---

[29] Augustine, *City of God*, XI:22; XII:3; XIV:11; *Enchiridion*, XIV.
[30] Tillich, *Systematic Theology*, 2:60.

The will, therefore, is then truly free, when it is not the slave of vices and sins."[31]

Two notions are attached to evil for Augustine: first, evil is the privation of what is good and has no eternal being in itself and, second, it is a temporary concession God has made to human history to allow the story of human freedom with its loss and redemption to go forward. Evil and its concomitant suffering are the price paid for a redeemed creation of freely loving citizens in the City of God. Evil is a negation that is negated by grace; it becomes a means for a higher redemption.

Augustine's theodicy is lumped with others making the free will defense because, according to this account, evil is the product of the creature's free decision. Even though God could foresee the human fall, God did not create evil. Evil is not a created entity, so it is the creation of no one. Yet, one can still assert that the responsibility for the fall into evil lies on human shoulders. God is responsible for providing us with the first freedom and the redeemed freedom by grace. God is justified by this theodicy, because God is not responsible for the origination of evil and yet is responsible for our redemption from evil.

Free will defenders contend that it is better for God to bring into being a creation which is capable of being corrupted, capable of having evil and suffering as a byproduct of free will, than it is to bring into being creatures without free will. With this axiom, evil becomes a tolerable risk. "In freely willing the other's freedom, God does come to be affected by the contingent reality of that which is not God" writes Paul Sponheim; "Evil deeds carry such efficacy as to weave a web of brokenness in which every human life is caught – that also speaks of the degree of risk and cost involved in God's will for the freedom of the other."[32] God here is further justified because freedom is such an overwhelming good that God rightly risked evil in order to make freedom possible.

In addition to the free will defense, other theodicies are available among contemporary theologians, most prominently the developmental view of John Hick following Irenaeus that sees evil and suffering as necessary to the growth process of soul-making.[33] Whiteheadian process theologians offer an alternative theodicy that denies the first proposition – they deny that God is omnipotent – while strongly affirming divine love and assigning God the role of luring actual entities beyond evil toward God's subjective aim.[34] All three of these positions – free

[31] Augustine, *City of God*, XIV: 11.

[32] Paul Sponheim, *Faith and the Other* (Minneapolis: Fortress, 1993) 91–92.

[33] John Hick, *Evil and the God of Love* (New York: Harper, 1978).

[34] Charles Hartshorne, *Omnipotence and Other Theological Mistakes* (New York: SUNY, 1984); Marjorie Hewitt Suchocki, *The End of Evil* (New York: SUNY, 1988).

will defense, soul-making teleology, and limiting God's power – affirm theism and do not yield to the conclusion that our experience of evil is sufficient to persuade us that no God exists. In addition, all three are not willing to give up on affirming God as gracious. God's love is not a negotiable item. If there is a God, then this God is loving, even all-loving. God is justified because the world's evil either lies beyond God's power to control or is taken up into a more inclusive good.

Is theodicy the best way into the question of God's action in the natural world? Is theodicy the way to pose the *Fragestellung*, to pose the question to be pursued? A summons to "justify God" ought to ring dissonantly in Reformation ears, especially Lutheran ears. For a theological tradition struggling to understand how God justifies us whose sins lead to evil and suffering, this reversal of the agenda ought to seem odd.

Just what is the barrier to grace here? It may seem like the evil we experience is the obstruction; yet, the cross, which takes sin and suffering unto itself, is the fundamental symbol for the presence of grace. So, this cannot be the barrier. Perhaps then the barrier is found in those philosophical reflections upon evil that nominate God as the one guilty for evil and, with a curious turn in logic, count this as evidence that God does not exist. The argument against God's existence includes the side corollary that if there would be a God who would permit such horrendous evils that this God would not be worthy of existence. In short, this is a moral argument, not an ontological argument, against God's existence. Defenders of God unanimously affirm in their theodicies that God is moral; God is all-loving despite the way the world looks.

I do not recommend we take on the argument in this form. We should alert ourselves to the dynamic of self-justification operative everywhere in human self-promotion; and we should look to see if it is operative even in theological discourse. Turning our attention in this direction uncovers a subtle assumption at work in both atheistic and theistic theodicies, namely, that our human perspective on evil and suffering is placed in the judgment seat. The theodicist has claimed the role of judge and placed God on trial. As Job found out, the mysterious and magnificent God who established the foundations of the world does not subject divinity to trial by humanity, even on moral charges (Job 38–42). Timothy Lull reminds us, "One thing we know about God, or ought to know about God, is that God is God and we are not."[35]

Perhaps this is what prompts Robert W. Jenson to sidestep the task of providing a theodicy within the scope of his systematic theology.

---

[35] Timothy Lull, "God and Suffering: A Fragment," *Dialog* 25:2 (Spring 1986) 94.

"No theodicy is proposed. It is in any case not possible within the system here presented to 'justify the ways of God to men'."[36]

Yet, an appeal to the privilege of divine mystery is but one resource Lutheran theology brings to the problem of evil. Again, the Theology of the Cross places itself front and center. Our concern in this instance is first an epistemological one, to be sure – the concern over knowing the hidden God within or behind the God revealed in the cross. The epistemological concern has to do with revelation. "Thus the characteristic of God as the *Deus theologicus* is that he is *Deus crucifixus* and *absconditus*," comments Gerhard Ebeling, indicating that to know God truly is to know God in the crucifixion.[37] This also has to do with avoiding a Theology of Glory (*theologia gloriae*), wherein we construct an image of God based upon the strengths and perfections we can imagine by looking at the works of creation. Such imaginings that seem to glorify God but in fact glorify our human imagination are just what get dowsed when turning our attention to the revelation of God in the cross. "No matter what the issue or the problem in life," writes Eric Gritsch, "the theologian must see it within the shadow of the cross. Thus theology must ultimately concentrate on Christ and not on the believer. True theology, therefore, always crucifies the theological ego."[38]

In addition to the epistemological problem with divine hiddenness, secondly, the Theology of the Cross also points to what is revealed, namely, that God is present in the suffering and death that comes with the cross. The crucifixion of Christ is God incarnate with us, taking evil and evil's guilt into the divine life. God is the victim of human evil, not its perpetrator; and God freely absorbs this victimage as an expression of a divine love that does not fight evil with evil but rather with the grace of forgiveness and reconciliation. Luther emphasizes "how much it cost Christ and what he paid and risked … he suffered, died, and was buried so that he might make satisfaction for me and pay what I owed, not with silver and gold but with his own precious blood."[39] One clear message relevant to the problem of evil is this: God is present with the world as it undergoes evil and present with us as we undergo

---

[36] Robert W. Jenson, *Systematic Theology* (2 Volumes: Oxford: Oxford University Press, 1997–1999) 2:21.

[37] Gerhard Ebeling, *Luther: An Introduction to his Thought* (Minneapolis: Fortress, 1970) 227. George L. Murphy is one of the few to creatively apply the Theology of the Cross to nature: *The Cosmos in the Light of the Cross* (Harrisburg: Trinity Press International, 2003).

[38] Eric W. Gritsch, *Martin Luther: God's Court Jester* (Minneapolis: Fortress, 1983) 168.

[39] Luther, "Large Catechism," in *The Book of Concord*, ed. by Robert Kolb/Timothy J. Wengert (Minneapolis: Fortress, 2000) 434–435.

suffering. God is not divorced from evil, but rather has become one with those of us who suffer evil. God is not immune to the suffering of the world, but rather shares in that very suffering.

The problem of evil is not reducible to a logical dilemma. Evil belongs to the drama of the world, a drama in which God struggles with and for the world. The resolution to the theodicy dilemma will not be found in clever redefinitions of the propositions or new insights into logical connections. Rather, we will find the meaning of evil and redemption from suffering in God's promise of a coming omega. By raising Christ from the dead on Easter, God has shown that the cross is not the last word. By promising an eschatological new creation, God has shown that the sins and sufferings of this present aeon are not eternal. Death will bring an end to suffering; and resurrection will inaugurate an aeon of healing. God's omnibenevolence is demonstrated by his empathic love in the cross; and God's omnipotence is demonstrated by his power over death, the power of creation and new creation. This is the drama in which we find ourselves, a drama with both evil and the promise of redemption from evil. Friends of God's grace share the anguish of those who suffer; they also share the promise of redemption.

Evangelical Ethics

Friends of God's grace also share the ethical commitment to combat sin and serve those who suffer. "For Luther, concern for the true knowledge of God and concern for the right ethical attitude are not separate and distinct but ultimately one and the same," says Paul Althaus making the move from the Theology of the Cross to ethics.[40] Tuomo Mannermaa makes the same move but launches it from the indwelling of Christ in faith; "Faith means participation in the being and thus in the properties of God. And one of the properties of which the Christian in his faith partakes is love. Christ, who is present in faith as *donum*, brings love with him, because Christ is in his divine nature God, and God is love."[41] Via either route, the destination is neighbor love.

Neighbor love (*Nächstenliebe*) – a love that requires definition according to the needs of the neighbor plus devotion to serve the neighbor as one's lord – has become a hallmark of evangelical ethics, especially in Lutheran circles.[42] The attention of faith is turned from our relationship to God toward the world. "God does not need our works,"

---

[40] Paul Althaus, *The Theology of Martin Luther* (Minneapolis: Fortress, 1966) 27.

[41] Mannermaa, "Why is Luther So Fascinating," 16.

[42] Luther, "The Freedom of a Christian," *Luther's Works*, 31:329–377.

says Marc Kolden, "our neighbors do."[43] Freed from the tyranny of judgment and death, the person of faith is freed to become voluntarily enslaved to the needs of neighbors who suffer. "The intimate connection in which Jesus places our relation to God and our relation to the neighbor presupposes that we are, as Luther expressed it, 'daily bread' in the life of one another."[44]

Conclusion

In our scientific and secular era, doubt and evil appear as barriers to experiencing and appreciating divine grace. Each denies the existence of God, doubt making a metaphysical argument and evil making a moral argument. Without God, we might find it difficult to make the claim that God is gracious. Yet, building a theological interpretation of reality is salutary only if it stands on a foundation of belief in a God who is gracious.

In addition, we must testify to a certain ambiguity. Both doubt and theodicy have some admirable qualities. Doubt, in the form of scientific agnosticism, retains an implicit commitment to truth. Evil, as reflected upon in theodicy, reminds us of our implicit commitment to a world that should be governed by love rather than suffering. What these barriers place before us requires respect. Theological construction should respect if not incorporate them.

More important than rendering apologetic proofs for the existence of God is proclaiming that the God in whom some believe and others don't is gracious. According to the Theology of the Cross, grace comes in the form of God suffering with us. This is also the God who redeems, who heals the broken, who lifts up the unworthy, who reconciles enemies, who invites the stranger into his eternal home. This is the only foundation worthy of further theological construction.

---

[43] Marc Kolden, "Ministry and Vocation for Clergy and Laity," in *Called and Ordained*, ed. by Todd Michol/Marc Kolden (Minneapolis: Fortress, 1990) 196.

[44] Knud Eijler Løgstrup, *The Ethical Demand* (Notre Dame: University of Notre Dame, 1997) 5.

Chapter 4

Grace in the Genes?

As heirs to our evolutionary history, what have our genes inherited? If we observe nature, we see the unavoidability of the predator-prey interaction and we see the extinction of ninety-five percent of all species thus far. We see what Charles Darwin had labeled "waste" in nature. That is what we observe. How do we explain what we observe? If we adopt the *selfish gene* theory of the sociobiologists and evolutionary psychologists, this waste is explained by the drive to replicate our DNA sequences. Not only nature's waste, but also all our human cultural glories are explained as selfish genes perpetuating themselves through our particular history of natural selection.

The selfish gene theory not only explains our survival on the grounds that our particular DNA sequences have proven themselves more reproductively fit, it also explains our propensity for violence. It explains why families stick together and why races and nations commit genocide against rival tribes. We have an apparent scientific explanation for what Christians have been referring to as original sin. "The roots of all evil can be seen in natural selection, and are expressed (along with much that is good) in human nature. The enemy of justice and decency does indeed lie in our genes," writes Robert Wright.[1]

Now, let us ask: can God's grace be mediated by our genes? Can God work through our physical inheritance?

This may seem like an odd question, because historically Christian spirituality has taken as its point of departure an assumed struggle between flesh and spirit. In this struggle, the spirit should win over the flesh. Now, we need to ask, are our genes siding with our flesh or our spirit? Or, are they neutral? Or, are some genes allies of flesh and others of spirit? Just how should we understand theological anthropology in light of modern science?[2]

---

[1] Robert Wright, *The Moral Animal* (New York: Pantheon Books, 1994) 151.

[2] This chapter draws on research leading to previously published pieces, "Can the Body Heal the Spirit?," in *Spiritual Information: 100 Perspectives*, ed. by Charles L. Harper, Jr. (Philadelphia: Templeton Foundation Press, 2005) 381–388 and "Holy Therapy" for *Christian Century*, 120:16 (9 August 2003) 23–26.

Before we proceed, we need to distinguish two kinds of genetics, the quasi-philosophical view of sociobiology and the empirical work of molecular biology. According to sociobiology's selfish gene theory, evolution is driven by DNA replication. It matters not at all which individual genes are involved. What matters only is the principle of replication through reproductive fitness. Research pursued by molecular biologists is quite different, because these scientists seek to uncover the expression and protein production of specific genes. If we ask about the possibility of genetic influences on human violence, sociobiology relies on a general principle of DNA replication whereas molecular biology searches for specific genes that might influence human behavior. In what follows, we will deal primarily with the molecular approach.[3]

The Struggle Between Flesh and Spirit

We find ourselves in a struggle between flesh and spirit, says St. Paul in the New Testament. When the Spirit wins, we enjoy the fruits: "love, joy, peace, patience, kindness, generosity, faithfulness, gentleness, and self-control" (Galatians 5:22). When the flesh takes control, it works "fornication, impurity, licentiousness, idolatry, sorcery, enmities, strife, jealousy, anger, quarrels, dissensions, factions, envy, drunkenness, carousing, and things like these" (Galatians 5:19–20). If this is the structure of Christian spiritual vision, then I would like to pose the following questions. First, can science aid us in understanding the influence or force of the genes on our fleshly or spiritual disposition? Second, can science and the technology it spawns help the spirit in its struggle with the flesh? If we are born with a genetic predisposition toward flesh, and if molecular biologists are able to influence gene expression through medical technology, might they guide gene expression toward more spiritual lives? If we devise a pharmaceutical to overcome an inherited genetic predisposition to antisocial behavior, for example, could such therapy be considered growth in the spirit?

Will Today's Science Reduce the Spirit to the Flesh?

Is the relationship between flesh and spirit the same as the relationship between body and soul? No. The struggle between flesh and spirit

---

[3] The alternative approach by examining sociobiology and its implications is also worthy of theological attention. See, for example, Uwe Gerber/Hubert Meisinger, ed., *Das Gen als Maß aller Menschen? Menschenbilder im Zeitalter der Gene* in *Darmstädter Theologische Beiträge zu Gegenwartsfragen* (Frankfurt a. M.: Peter Lang, 2004).

ought not to be equated simply with the distinction between body and soul, even though many have lived vibrantly with this partially misleading equation.[4] Debates go unabated as to whether the New Testament holds to a metaphysical dualism between body and soul. Rather than risk begging the question by taking a position in advance, in what follows we will stick to our formulation of the second question – "can science aid the spirit?" – and draw implications from that point. Here we merely note in a preliminary way that in St. Paul's theology, both body and soul are caught up in the war between flesh and spirit. Flesh and spirit are forces exerted on both body and soul, even if it seems at first glance that flesh wells up out of the body to contaminate the soul. By "flesh" we mean the body at war with the Spirit of God; and fleshly forces can dim the soul and coerce the soul into doing their bidding. The countervailing force of the divine Spirit liberates the self, both body and soul, from fleshly degradation.

Despite the above clarification, Stoics and ancient Christian theologians have tended to associate flesh with body and spirit with soul. They dubbed the former our lower nature and the latter our higher nature. The spiritual task, they said, was to cultivate the life of the mind or soul so that our higher nature could gain control over our lower nature. We are born dominated by our body; we should die victoriously free in the life of the liberated mind. Our higher natures cleave to sublime thoughts about God, to virtuous ideals such as justice and beauty, to moral integrity, and to mental disciplines such as prayer and meditation. These higher thoughts are eternal, whereas thoughts about cravings dictated by the body are ephemeral. Strength in the soul enables the higher or spiritual dimension to gain freedom from our bodies through control over the source of fleshly desire. Whether they used the word 'soul' or 'mind' they were actually seeking victory for the spirit. The result of such thinking led to centuries of spiritual practices where the goal was to gain mental control over the physical, immaterial control over the material, and hence spiritual control over the flesh.

This history of spirituality provides a valuable resource for assessing the rapid pace of developing knowledge about human nature rising out of research in molecular biology and the neurosciences. Although to date the experiments have been few and empirically reliable information remains fractional at best, a mood of biological determinism is be-

[4] Perhaps the most articulate of scholars dealing with questions surrounding substance dualism is Nancey Murphy. See her work plus that of her colleagues in Warren S. Brown/Nancey Murphy/H. Newton Malony, ed., *Whatever Happened to the Soul?* (Minneapolis: Fortress Press, 1998) and Ted Peters/Robert John Russell/Michael Welker, *Resurrection: Theological and Scientific Assessments* (Grand Rapids: Wm. B. Eerdmans, 2002).

ginning to emerge. Researchers are looking for genetic predispositions to mental states and propensities toward anti-social behavior; and brain researchers are looking for roots to intellectual and emotional processes in neuronal activity.[5] The specters of genetic determinism and biological reductionism are looming on the horizon. Materialist philosophers are employing such science to justify their eliminative reductionism, according to which no soul exists, or spirit either, for that matter. Both the dialectic of spirit-flesh and soul-body are under siege.

Is contemporary science about to take away our freedom right along with our souls and spirits? Are biologists about to imprison the human spirit in a fleshly jail? Is our higher nature reducible to our lower nature, so that the spirit no longer has leverage to deal with our fleshly desires? The level of alarm need not be shrill, I believe. There is no need to react like Pavlov's dogs in defense of human freedom against the alleged scientific jailers. There is no need to avoid the science on the grounds that it is irrelevant to truths already firmly and independently established by religion.

Given what we have just noted regarding the tradition of theological thinking about the struggle between flesh and spirit, religious reactions to biological determinism are likely to be plastic and protean and even creative. They will survive the culture war. But, how?

Genes, Nurture, and Anti-Social Behavior

Here we will take a brief look at a single scientific study, the kind of study we would expect on the part of those seeking biological explanations for human behavior. Under the hypothesis of genetic determinism or, perhaps more modestly, genetic influence, questions are being asked about biological factors in human behavior. This becomes especially relevant to theology when the behavior in question is either sinful or virtuous. Do genes make us sin? If so, does this contradict or complement what Christian theologians have previously thought about our bodily inheritance and its influence on the soul, or even on the spiritual dimension of who we are?

We will nickname our sample study the *X chromosome study*, because it addresses the question of genetic influence in moderating environmental influence on human behavior, specifically, antisocial or

[5] See: Ted Peters, *Playing God? Genetic Determinism and Human Freedom* (London /New York: Routledge, [2]2002) and Robert John Russell/Nancey Murphy/Theo C. Meyering/Michael A. Arbid, ed., *Neuroscience and the Person* (Vatican City State/Berkeley: Vatican Observatory and CTNS, 1999).

criminal behavior. The study's conclusion is that if a certain gene regulator on the X chromosome permits only a low level of expression of the monoamine oxidase A gene, then young boys have an increased risk of going to jail; conversely, a higher level of this gene's expression reduces the risk of engaging in the kind of antisocial behavior that makes a young man jail bound.

The X chromosome study looked at young boys, maltreated in their youth, whom we would expect to grow up to become criminals. Yet, an identifiable level of a certain gene's activity seems to reduce the otherwise fatalistic power of a family environment of abuse.[6]

Here is the research question: why do some male children who are maltreated in their home grow up to develop antisocial behavior traits while others do not? The assumptions orienting the research question are worth noting. First, male children rather than female children were selected; because the researchers were already looking for a factor on the X chromosome which only the male gender carries – that is, they assumed that the antisocial behavior in question is a gender specific phenomenon. Second, the researchers assumed that maltreatment of young boys increases the risk that they will grow up exhibiting antisocial personality symptoms, and conduct disorder, and becoming violent offenders – that is, they assumed that a social environment of victimization exerts a strong influence toward becoming a victimizer.

Building upon these assumptions, the researchers constructed their experiment on top of a previous study of 1000 children, half boys and half girls, who were assessed for maltreatment at ages 3, 5, 7, 9, 11, 13, 15, 18, 21 and 26. Of these 8 % were categorized as having experienced "severe maltreatment" between years 3 and 11; 28 % were dubbed as "probable maltreatment" cases; and 64 % experienced no maltreatment. The researchers focused on the first category, 26 year old males who had been severely maltreated between the ages of 3 and 11, and slated them for genetic testing. Then on the X chromosome they examined the gene for monoamine oxidase A (MAOA), a gene that governs a neurotransmitter-metabolizing enzyme in the brain. Actually, they examined the promoter, or the regulatory sequence for this gene expression, finding a polymorphism that down regulates gene activity. Those young men whose MAOA gene exhibited low expression levels were much more likely to exhibit aggressive antisocial behavior and become incarcerated for violent crimes than those whose gene exhibited a high level of expression.

---

[6] Avshalom Caspl,/Joseph McClay/Terrie E. Moffitt/Jonathan Mill/Judy Martin/Ian W. Craig/Alan Taylor/Richie Poulton, "Role of Genotype in the Cycle of Violence in Maltreated Children," *Science*, 297 (2 August 2002) 851–854.

The research team noted that the effect of childhood maltreatment on antisocial behavior was significantly weaker among males with high MAOA activity. More broadly, they noted that maltreated males with low MAOA activity genotype were more likely than nonmaltreated males with this genotype to be convicted of a violent crime by a significant ratio, thereby, reinforcing the environmental assumption identified above. Finally, the scientists concluded that the association between maltreatment and antisocial behavior is conditional, depending on a child's MAOA genotype. In sum, environmental or social influences are relevant but insufficient to explain antisocial behavior; genotype must be factored in. The DNA is decisive.

As we tease out the significance of this X chromosome research, we note that both assumptions and conclusions are deterministic in structure. They begin with the assumption of environmental determinism – namely, if maltreated then young boys will grow up antisocial – and then shift to genetic determinism – gene expression exacerbates or mitigates environmental influence. The net effect of both the assumptions and the conclusions is that some boys are born into situations in which the combination of gene expression and social context heavily determine what kind of person they will be in adult society.

Are We Born Morally Neutral?

Now, let us ask: do such findings by contemporary science contradict or complement what theologians have traditionally believed? We could imagine that a modern Pelagian might want to defy the science of the X chromosome study by asserting that we are born morally neutral, that we enter the world and grow up with the capacity to decide equally between right and wrong. Pelagius himself in the fifth century actually showed little interest in the concept of original sin. His thesis was that each human being could take initiative toward salvation from our natural state without the aid of divine grace.

Yet the Pelagian tradition down to modern humanism emphasizes original neutrality over against sin. Good and evil here are thought to be equal options standing before a freely deciding human psyche. So, assumptions about determinism, either biological or social, would have to be dismissed as compromising this morally neutral anthropology. Even a softer or less deterministic interpretation such as we find characteristic of molecular biology, according to which gene expression predisposes or inclines a person in one direction or another, would obviate the classical Pelagian assumption that we are born morally neutral. Might science be mistaken if it asserts that some of us if not all of us are destined by our genes to be predisposed toward evil behavior? The theological position that we are born morally neutral will find rough sledding in this scientific environment.

A follower of Augustine, in contrast, might see such scientific research as partially demonstrating what Christians have known all along, namely, we emerge from our mother's womb with a self-orientation that makes loving God and loving neighbor contradictory to our innate predisposition. We are born *homo incurvatus in se*, curved in upon ourselves. It takes an act of divine grace to reorient us toward loving God and loving our neighbor as we would love ourselves. It takes the inspiration of the Holy Spirit to orient our hearts and wills and minds toward expressing the fruits of the Spirit.

So, what we have here in theology is much more broad and sweeping in scope than what appears in scientific study, obviously. Such research does not even ask about the total orientation of the human self. It deals only with one segment of human behavior, and a pattern of behavior that applies to some but not all of us. Does this obviate the value of comparing science and theology? By no means. Such science is still quite relevant to theological anthropology. If genetic inheritance and social inheritance combine to predispose us to behavior with moral valence in some cases, then we can hypothesize that some level of biological and environmental determinism has an effect on everyone's life. Our genes and our family experience provide both opportunity and constraint for the kind of person we will grow up to be.

Although we are focusing here on the predisposition toward sinful behavior, we could prognosticate that in another setting we might provide a parallel analysis of caring behavior. I believe we can safely assume that favorable genotypes and certainly loving family contexts increase a child's opportunity to grow up with high minded values and an increased capacity for loving his or her neighbor.

Further still, a nuance related to the X chromosome study might be worth pondering here. The young men studied were victims of maltreatment. We might wish to ask: do they love themselves? Does their whelming experience of abuse permit the emergence of self-love, or might maltreatment more likely retard the growth of self-worth and leave the child with self-loathing? Might the antisocial behavior in question be an expression of self-hatred rather than a self-love unable to expand to include others? If the Augustinian lens through which sin is interpreted is this – sin is too much love for self and not enough love for God or neighbor – then perhaps we need a more subtle analysis of the young men in the X chromosome study. If scripture is right that we love because God first loved us (1 John 4:19), then perhaps all of us, these young men included, need first to experience love before the capacity to love either self or neighbor can develop. Perhaps we need to experience unconditional love before we can develop the capacity to love others unconditionally. This may be the way grace works in a redemptive way.

Original Sin as Inherited Sin

Talking about *original sin* creates discordant sounds in the ears of modern intellectuals. The concept is unwelcome, even shunned. Perhaps this shunning is due to the historical connotations of the term 'original'. The picture painted by Augustine is that Adam and Eve back in the Garden of Eden committed the first sin, the original sin; and through procreation they have passed this fallen state on to each subsequent generation. We all inherit Adam's sin; we all participate in Adam's sin.

Two contemporary objections pop up. First, our modern notions of justice would limit our responsibility to our own sins; we should not be held accountable for an action of someone else such as Adam or Eve. The second objection is that the Augustinian history is apparently no longer acceptable in a Darwinian era. The dominance of the theory of evolution with its deep time and epic of human emergence from previous species has no room for a myth of origin that places the human race in a prior state of grace. The Garden of Eden cannot be located geographically or geologically. Rather than a fall from a pristine state, modern science sees the human race arising from a long struggle characterized by natural selection and survival of the fittest.

The concept of 'original' in 'original sin', it seems to me, does not require a history that includes a past Garden of Eden with a now lost perfection; nor does it require blaming Adam and Eve for our own moral condition. Rather, it is sufficient for 'original' to refer to the 'origin' of each one of us. Our own individual origin is characterized by conditioning, by genetic conditioning and family context conditioning. We are born with opportunities and constraints over which we had no original control, and some of this conditioning influences our predisposition to behavior toward others.

This leads me to ask: should we try to retrieve here the theological term *inherited sin*? When the Augsburg Confession discusses "Original Sin" we read that all of us who are conceived according to nature are born into sin, that we are "full of evil lust and inclinations" from our mother's wombs onward; and we are unable by nature to have true fear of God and true faith in God. This state is referred to as an "inborn sickness and hereditary sin (*Erbsünde*)."[7] John Calvin similarly

[7] *Augsburg Confession*, Article III.

interprets the concept of original sin so that what we inherit becomes prominent. "Original sin, therefore, seems to be a hereditary depravity and corruption of our nature, diffused into all parts of the soul, which first makes us liable to God's wrath, then also brings forth in us those works which Scripture calls 'works of the flesh'."[8]

Much is being said here. The single item I would like to lift up is the ease with which such theological thinking accepts the notion of inheritance. When we are born we find ourselves already conditioned, already predisposed toward a life that is alienated from God's will, even alienated from faith in God. Although the presumption here is that this inheritance is physical – that is, passed on through conception and birth – the point is that we begin our life of moral responsibility already conditioned by factors beyond our control. It is certainly consistent, then, for Social Gospel theologians such as Walter Rauschenbusch early last century to observe how prejudice and social discrimination is passed down from one generation to the next; and it is consistent for liberal theologians today to incorporate observations about social inheritance – what liberation theologians and feminist theologians call 'social location' or 'systemic evil' – into our understanding of the human condition. Whether biological or social, whether innate or environmental, we begin our morally responsible life with a specific inheritance and a predisposition toward behaving in specific ways.

Fruits of the Spirit

Now, let us try on for size a speculative scenario. Suppose a clever medical scientist could invent a pharmaceutical capable of up regulating the expression of monoamine oxidase A. Suppose the young boys subjected to family abuse could have access to MAOA therapy; and suppose that this genetic therapy would strengthen the influence of genetic determinants over environmental determinants. Suppose the result would be that when attaining adulthood these young men would possess a greater sense of social responsibility exercised through greater self-control. Theologically speaking, should we consider this to be a fruit of the Holy Spirit?[9]

---

[8] John Calvin, *Institutes of the Christian Religion*, II:i:8.

[9] "Recent genetic studies of behaviour have begun to dissolve the distinction between matter and spirit," avers University of Virginia genetics researcher, Lindon B. Eaves, "'Ought' in a World that Just 'is'," *Is Nature Ever Evil?*, ed. by Willem B. Drees (London/New York: Routledge, 2003) 284.

To this I would offer a qualified "yes" answer. As Augustine sorts out the dialectic of sin and grace, he begins with bondage to sin and then moves to liberty from it. "He who is the servant of sin is free to sin. And, hence, he will not be free to do right, until, being freed from sin, he shall begin to be the servant of righteousness. And this is true liberty."[10] Might we think of the combination of genetic predisposition and maltreatment in youth as a form of bondage; and might we think of medical therapy that readies a person for increased self-control a form of liberation?

Note a pitfall I am trying to avoid here. I am trying to avoid defining the question in terms of metaphysics. If we are assuming that the distinction between body and soul or the conflict between flesh and spirit are metaphysical divisions, then the situation would be conceptually hopeless. If we would assume that body and soul are different substances and that flesh and spirit are metaphysical opposites, then the spiritual conflict precipitated by gene function would be unanalyzable. However, if we presume that a person is a psychosomatic unity, that who we are as persons is inclusive of both body and soul, then their interaction at the level of the human self becomes accessible.

Recall that earlier I suggested we can best understand flesh and spirit as forces, not as merely alternative terms for body and soul. One of the fruits of the spirit in Saint Paul's list is "self-control." No matter what genotype we are born with or what family pattern of rearing we experience, self-control remains an achievement that each self must attain in the maturing process. In common parlance, 'self-control' as a term means what it says, namely, the self takes over a level of control previously under the hegemony of bodily cravings and social influences. If a pharmaceutical could enhance one's capacity for self-control, such therapy could very well be thought of as a spiritual force. That such therapy works on the body does not make it any less spiritual. Nor does it make medical therapy anything less than an expression of God's grace in the life of a person who benefits from it.

Need we who are theologians become anxious that modern medicine will put us out of a job? No worry is fitting here. The struggle between flesh and spirit is a big one; and winning one little battle over genetic expression of MAOA does not in itself indicate that we are ready to declare total victory in the war against the flesh. The Spirit's orchard covers many acres; and there are many more fruits of empowering grace the Holy Spirit can cultivate through either our body or our soul.

---

[10] Augustine, *Enchiridion*, XXX.

Chapter 5

The Hermeneutics of Evolution

That which makes modern Christian theology distinctively modern is the focal role played by the hermeneutical question. Not every theologian states the hermeneutical question directly or succinctly; yet, this question has guided nearly every school of modern theology since Friedrich Schleiermacher published his *Speeches on Religion* in 1799. The question has to do with our own hearing: how can we hear the biblical Word of God with ears that interpret everything through modern consciousness?[1] All hearing is interpretive; and we modern people interpret everything we experience in light of a worldview or world picture projected by natural science. When Karl Rahner takes up the matter of evolution, he places it within the question "concerning the reconciliation of the Christian teaching and interpretation of evidence with the present-day way of living, thinking, and feeling."[2]

The hermeneutical question asks how theology with ancient roots can flower in a modern and emerging postmodern world. For theology to be understandable and credible it must explain just how it relates to a worldview shot through and through with the scientific interpretation of nature. Theology need not be reduced to scientific categories, to be sure; yet the categories of the theologian need to be understandable in light of science. And, to date, the discipline within the sciences posing the greatest challenge to a theological understanding of the human condition and of God's action in the natural world is evolutionary biology. Since the era of Charles Darwin, it has appeared that evolutionary theory would supplant rather than comple-

---

[1] "*Our own hearing is required. That is the hermeneutical problem in our daily life.*" Ernst Fuchs, "The New Testament and the Hermeneutical Problem," in *The New Hermeneutic*, ed. by James M. Robinson and John B. Cobb, Jr. (San Francisco: Harper, 1964, italics in original). Hermeneutics as a discipline is not limited to the practical task of interpreting; it also includes the fundamental investigation into human understanding and theories of interpretation. See: Werner G. Jeanrond, "Hermeneutik, "V. Fundamentaltheologisch," RGG^4, 3:1654–1659.

[2] Karl Rahner, "Christology within an Evolutionary View of the World," *Theological Investigations* 22 Volumes (London: Darton, Longman, and Todd, 1961–1976; New York: Seabury, 1974–1976; New York: Crossroad, 1976–1988) V:159.

ment the Christian vision of reality. Just how should contemporary theology deal with modern evolutionary theory?

The hermeneutical problem is exacerbated because the modern world picture based upon natural science has closed all windows to transcendence. Methodologically, scientists search solely for natural or material explanations of phenomena. When this methodological commitment drifts over into ontology, then we draw a picture of the natural that shuts out not only the supernatural but every other transcendent dimension. Ontological materialism and atheism are the result.

In what follows we will ask the hermeneutical question. We will formulate it in such a way that we can search for signs of transcendence within the horizon of secular experience, especially within the horizon of the evolutionary worldview. We will look at a selection of theological answers, each one of which attempts to keep the windows to transcendence open. We will look briefly at the work of Langdon Gilkey, Huston Smith, the Intelligent Design school of thought, and Theistic Evolution. Our analysis will include an examination of the *naturalistic fallacy* in Social Darwinism and Sociobiology. The task of a hermeneutics of evolution is to keep the windows of transcendence open while affirming the new discoveries about our natural world that progressive science offers.

The Hermeneutical Question

Pursuing possible compatibility between theology and evolutionary theory is one approach to answering the hermeneutical question. I formulate the hermeneutical question this way: "*how can the Christian faith, first experienced and symbolically articulated in an ancient culture now long out of date, speak meaningfully to human existence today as we experience it amid a worldview dominated by natural science, secular self-understanding, and the worldwide cry for freedom?*"[3] Among the characteristics of the modern world since the Enlightenment is a worldview dominated by natural science. The question is this: how should we interpret the Christian faith in light of this worldview? This does not necessarily require translating or recasting or rationalizing or mathematizing the Christian faith to make it fit a scientific worldview; rather, it simply asks: what ought we as Christians to do in this context?

[3] Ted Peters, *GOD – The World's Future: Systematic Theology for a New Era* (Minneapolis: Fortress Press, [2]2000) 7, italics in original.

Is modernity enough? What about postmodernity? One of the characteristics of our contemporary culture – the emerging postmodern culture – is the presence of plurality and pluralism. *Plurality* refers to the diversity of perspectives and social locations of interpreters; *pluralism* is the ideology that incorporates plurality into its methodology demanding that all voices regardless of social location be heard. The hermeneutical question has been expanded by theologians to include postmodern pluralism with its accompanying ambiguity regarding truth claims. "All experience and all understanding is hermeneutical," writes David Tracy adding that "postmodernity demands multiple discussions for interpretation itself."[4]

This has significance for the conversation between theology and science; because science is both modern and universal in its claim to be the guardian of objective knowledge. Should theology interpret the ancient gospel strictly in terms of the worldview fostered by modern science, it would adapt to modernity but miss postmodernity. "No one can escape the problems of pluralism and discover 'universality' and 'objectivity' simply by appealing to scientific method," warns William Placher.[5]

Actually, plurality and ambiguity are not the only forces in emerging postmodern consciousness. Simultaneously at work is a contradictory force, namely, a yearning for holism, a cultural thirst for integration and healing. Whereas pluralistic postmodernism divides and fragments, holistic postmodernism unites and heals. Theologians seeking to answer the hermeneutical question need to take both cultural forces into account. So, I would like to offer a postmodern amendment to the hermeneutical question: "*how can the Christian faith be made intelligible amid an emerging postmodern consciousness that, although driven by a thirst for both individual and cosmic wholeness, still affirms and extends such modern themes as evolutionary progress, future consciousness, and individual freedom?*"[6]

When we turn to the evolution controversy, we can take partial advantage of pluralism in postmodern thought. To do so, we need to distinguish between plurality of social locations and the plurality of considered judgments. Less relevant is the alleged determinism of social location; more relevant is the difference in judgment and opinion that lead conscientious persons to dispute with one another. Theologians

[4] David Tracy, *Plurality and Ambiguity: Hermeneutics, Religion, Hope* (San Francisco: Harper, 1987) 77–78.

[5] William C. Placher, *Unapologetic Theology: A Christian Voice in a Pluralistic Conversation* (Louisville KY: Westminster John Knox Press, 1989) 14–15.

[6] Peters, *GOD – The World's Future*, 8, italics in original.

who share the same social location can disagree vehemently regarding what position to take on the relation of evolutionary biology to Christian faith. What they all share in common is the common question, the hermeneutical question regarding the meaningfulness of the Word of God within the context of scientific understanding. Richard Carlson asks it this way: "*How is Christian faith related to developments in contemporary natural science, especially to biology, earth science, and cosmology?*"[7] This is the question. What is the answer? Carlson recognizes that theologians with equal intellectual integrity can arrive at differing answers to this question. "I am convinced that there is no *single distinctly Christian viewpoint* on matters of the relationship of natural science and Christian faith. There are, however, *distinct viewpoints held by Christians ...*"[8] The plea here is that theologians respect one another in their differences.

The Hermeneutic of Secular Experience

The late University of Chicago theologian, Langdon Gilkey, adds an important ingredient to our hermeneutic of evolution. Gilkey called it a "hermeneutic of secular experience." This means the theologian provides a critical analysis of secular claims to discern where windows to transcendence might have been unnecessarily boarded up. Human experience is generally open to transcendence; so when scientific methods close that opening, someone needs to point it out. "We are trying to conduct a *hermeneutic of secular experience*," writes Gilkey, "to see what religious dimensions there may be" within secular self-understanding.[9]

Science, especially evolutionary biology, boards up the windows to transcendence when it undergirds ideologies such as Social Darwinism or atheism. As an ideology, the scientific worldview purports to answer questions of ultimacy. At this point, science oversteps its bounds. When science tries to establish a worldview, it has exceeded is limit. Scientists become dangerous when they try to "explicate a vision of the ultimate nature of reality or of process as a whole, and seek to understand man's nature, obligations, and destiny in the light of that total cosmic vision."[10] When this happens theologians need to

---

[7] Richard F. Carlson, ed., *Science and Christianity: Four Views* (Downers Grove IL: InterVarsity Press, 2000) 12, italics in original.

[8] Ibid., 11, italics in original.

[9] Langdon Gilkey, *Naming the Whirldwind: The Renewal of God-Language* (Indianapolis/New York: Bobbs Merrill Company, 1969) 234.

[10] Langdon Gilkey, *Religion and the Scientific Future* (New York: Harper, 1970) 73.

become prophets, and render judgment against the scientific usurpation of culture.

With regard to methodology, the early Gilkey embraced the two language or independence view of the relationship between theology and science. Near the end of his career he moved closer to dialogue or mutual interaction. When employing the hermeneutic of secular experience, Gilkey limited the domain or language of science to objective discourse about the mundane world. Science asks, "how?" Religion asks, "why?" When it comes to questions about the origin of the universe, science asks about *proximate* origins, whereas theologians ask about *ultimate* origins. By speaking two languages, these need not get confused. We confront problems only when scientists try to answer questions of human meaning, or when theologians try to assert facts. If each would recognize its own domain of speech, we would have peace rather than warfare betweens science and religion.

Gilkey seldom dealt with the evolution controversy directly. In the case of Arkansas Law 590, however, Gilkey served as a court witness. The problem with the Scientific Creationists, complained Gilkey when on the witness stand in Little Rock, Arkansas, in December 1981, is that they confuse the two languages. The state of Arkansas should not require the public schools to teach the story of creation as the Bible tells it within science classes, Gilkey told the court, because this confuses science and religion. "If, then, the kinds of truth in science and in religion are distinguished, there is essentially no threat of religion to science or of science to religion."[11]

A decade later, Gilkey began to move regarding his assumption that science and religion should be so separated. Gilkey became exposed to the "hypothetical consonance" and "creative mutual interaction" models for relating science and theology. In his later years, he began to take short steps in this direction. "Science and religion are mutually interdependent," he wrote; "the issues of the truth of science and the truth of religion and of the relations between these sorts of truth represent fundamental concerns for each."[12] Gilkey saw the need for greater interaction because of the question of truth claims. Yet, Gilkey feared that too much coziness between theologians and scientists might silence the prophetic voice the theologian must raise when science usurps culture.

---

[11] Langdon Gilkey, *Creationism on Trial: Evolution and God at Little Rock* (New York: Harper, 1985) 53.

[12] Langdon Gilkey, *Nature, Reality, and the Sacred: The Nexus of Science and Religion* (Minneapolis: Fortress Press, 1993) 11.

From Science to Ideology

I would like to distinguish between evolutionary biology as a scientific research program from evolution as an ideology. As a scientific theory, Darwin's principles of random variation and natural selection have been fertile. Especially when combined with genetics to establish the neo-Darwinian synthesis, evolutionary theory has not only predicted what we might find in fossil remains but has contributed to predictions about genetic mutations and provided what virologists need to devise new medicines. As science, the Darwinian model of evolution is solid.[13]

The late Stephen Jay Gould searched for the essence of the Darwinian model or theory and found three hard core principles: (1) *agency* is ascribed virtually exceptionlessly to organisms acting; it is a single-level theory; (2) *efficacy* in speciation is ascribed to natural selection and the slow accumulation of positive effects of favorable variations through innumerable generations; (3) the *scope* of this theory applies to all species over deep time, an extrapolation from microevolutionary processes and small incremental changes that have lead to the panoply of taxonomic diversity we witness today.[14] This tripod has produced a fertile scientific theory that continues to lead to new knowledge in many fields.

Yet, Darwinism as a scientific horse has never been without ideological riders. Take for example, the doctrine of progress. By the time Charles Darwin appeared on the scene in the middle of the nineteenth century, the idea that the human race was progressing in both technology and politics was firmly established. Would Darwin's evolutionary horse carry progress to new places? Would just plain change over time become change toward betterment?

Darwin himself affirmed at least a loose connection between evolution and progress; and his disciples who believed in social progress could then ground their belief in biology, in nature. According to Darwin, the law of progress is not an invariable rule; but in general we can expect evolution to move life toward approximating perfection. Darwin writes, "As natural selection works solely by and for the good of each being, all corporeal and mental endowments will tend

---

[13] The American Association for the Advancement of Science declared evolution to be the breakthrough theory of the year 2005. Giant leaps forward in genetics and speciation demonstrated "evolution in action." The point is that the theory generates new research the yields new knowledge. Elizabeth Cultotta/Elizabeth Pennisi, "Evolution in Action," *Science*, 310:5756 (23 December 2005) 1878–1879.

[14] Stephen Jay Gould, *The Structure of Evolutionary Theory* (Cambridge MA: Harvard University Press, 2002) 14–15.

to progress toward perfection."[15] While most of Darwin's disciples wallow in the doctrine of progress; some sharply repudiate it. The nay-sayers deny progress in nature; and they even deny that Darwin himself affirmed progress. Daniel C. Dennett, for example, writes, "Global, long-term progress, amounting to the view that things in the biosphere are, in general, getting better and better and better, was denied by Darwin, and although it is often imagined by onlookers to be an implication of evolution, it is simply a mistake – a mistake no orthodox Darwinians fall for."[16] What this shows is that we can distinguish between Darwin's science and the ideology of progress, even if we have to excise or abstract the science from Darwin himself. Regardless of who wins this debate, I wish to point out that the ideology of progress has been trying to ride Darwin's scientific horse since the origin of *Origin*.

Progress is not the only rider. Let me offer a slightly more comprehensive list. Already in Charles Darwin's own lifetime four potent ideologies cracked the Darwinian whip: (1) the Social Darwinism of Herbert Spencer relied on natural selection, referred to as "survival-of-the-fittest", to support a naturalistic ethic of *laissez faire* capitalism;[17] (2) the Eugenics of Francis Galton, Darwin's cousin, said we could speed up evolutionary progress through social programs to improve human heredity; and Adolph Hitler sought to speed up evolution with his program of racial hygiene that led to genocide;[18] (3) the atheistic tradition beginning with Thomas Huxley employed the methodological naturalism of Darwin to support what we now know as ontological materialism;[19] and (4) the already pervasive concept of progress previously confirmed by the advances of the industrial revolution received from Darwinian theory the added support that

[15] Charles Darwin, *On the Origin of Species* (London: John Murray, 1859; reprinted Cambridge MA: Harvard University Press, 1964) 489.

[16] Daniel C. Dennett, *Darwin's Dangerous Idea: Evolution and the Meanings of Life* (New York: Simopn & Schuster, 1995) 299.

[17] See: Herbert Spencer, *Social Statistics* (New York: D. Appleton, 1864).

[18] See: Francis Galton, *Hereditary Genius: An Inquiry into its Laws and Consequences* (London: Macmillan, 1892) and the new comprehensive treatment by Nicholas Wright Gillham, *A Life of Sir Francis Galton: From African Exploration to the Birth of Eugenics* (Oxford/New York: Oxford University Press, 2001).

[19] Thomas Huxley, while relying on evolutionary theory to support his anti-church materialism, pitted modern liberal ethics against the precedents set by survival-of-the-fittest. "Ethical nature, while born of cosmic nature, is necessarily at enmity with its parent ... The conscience of man revolts against the moral indifference of nature." *Evolution and Ethics* (London/New York: Macmillan, 1893) viii, 182. Huxley did not endorse the morally complacent evolutionary ethic of Herbert Spencer. It is the Spencerian form evolutionary ethics which dominated and became the prompt for discussions of the naturalistic fallacy.

progress is built into our biological substrate.[20] These are the ideological overlays requiring a hermeneutic of secular experience that is prophetic in force.

When we confront evolution at the cultural level, it comes to us already shrink-wrapped in ideology. We actually have to abstract the research science from its cultural packaging. The hermeneutic of secular experience both unwraps the science while examining and interpreting the shrink-wrapping.

Progress is the most common label on the ideological wrapping. New Age Spirituality, for example, has digested progress so thoroughly that the idea of evolution now applies to spiritual growth, even cosmic spiritual growth. The concept of "conscious evolution" based on the work of Teilhard de Chardin and Sri Aurobindo seeks to integrate the physical and the spiritual. The Foundation for Conscious Evolution puts it this way: "*Conscious Evolution* as a new world view is a context for a meta-discipline composed of the growing edge of all disciplines seen as parts of a living whole. It may even set the framework for a 'meta-religion,' an all-inclusive space in which the great faiths and wisdom traditions can contribute their unique gifts to the conscious evolution of humanity. The promise of *Conscious Evolution* is nothing less than the emergence of a universal humanity capable of co-evolution with nature and co-creation with Spirit."[21]

Such New Age versions of contemporary evolutionary ideology are not commonly embraced by laboratory scientists. Sociobiology is. In what follows we will look briefly at sociobiology and its century old predecessor, Social Darwinism. We will begin by looking at what connects them, namely, the naturalistic fallacy.

Let us compare the science resulting from Darwin's fundamental insights into random variation and natural selection to a petroleum deposit. It is a valuable resource and, thereby, a most valuable deposit. Yet, to tap it we must drill deeply to penetrate the layers of ideology that cover it up. The legacy of nineteenth century Social Darwinism inherited by twenty-first century sociobiology is the overlay to which I will give some attention in what follows.

---

[20] Enthusiasts for evolution "to a person were supporters of the a notion or philosophy of progress – the belief that through human effort we can improve our lot and that of our fellows – and they used the idea of biological progress (as they interpreted evolution) as an illustration and support of this idea." Michael Ruse, "Evolutionary Ethics Past and Present" in *Evolution and Ethics*, 27.

[21] http://www.evolve.org/pub/doc/evolve_what_is_ce.html.

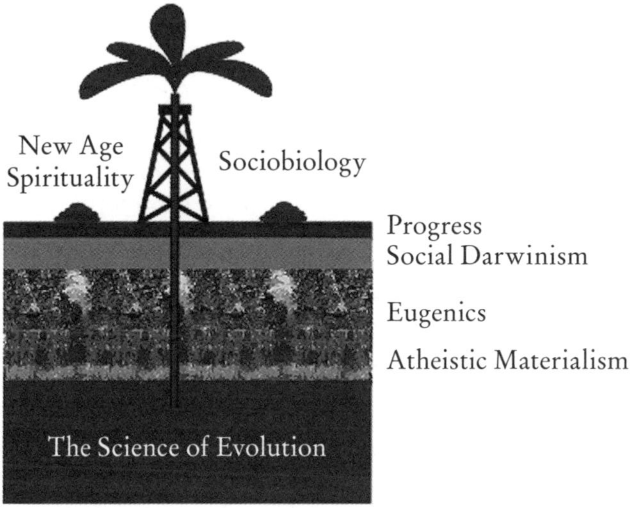

Figure 1: Evolutionary Ideologies by Martinez Hewlett. Used with permission.

Evolutionary Ethics and the Naturalistic Fallacy

The idea of the survival-of-the-fittest led to the rise of evolutionary ethics and the justification on the part of Social Darwinists for disregarding social responsibility to care for the poor, the infirm, and the economically weak among us. According to Herbert Spencer, evolution teaches us how to handle the poor in society. "The whole effort of nature is to get rid of such, to clear the world of them and to make room fore better ... it is best that they should die."[22]

Nature as described by science provided the grounding for evolutionary ethics and social disregard for caring. A hermeneutic of secular experience shows, however, that such an ethical vision is fallacious. It is based on an attempt to use science to close the windows to transcendence. To the naturalistic fallacy we now turn.

By the naturalistic fallacy we refer to the logic of basing what *ought* to be on what *is* in nature. From Social Darwinism we inherited a belief that progress can be found within nature, that to be more highly evolved is to embody that progress, and that fostering survival of the ever more fit is an ethical goal sanctified by nature's blessings. Charles

[22] Spencer, *Social Statistics,* 414–415

Sanders Pierce recognized the influence of *laissez faire* capitalistic theory on Charles Darwin's theory of competition between species for survival, saying Darwin "merely extends politico-economical views of progress to the entire realm of animal and vegetable life ... Every individual for himself, and the Devil take the hindmost."[23]

A century later, Social Darwinism is back. Late twentieth century sociobiology calls us to a reverence for reproductive fitness in the complex form of reciprocal altruism, a pattern devised by pre-human communities that sets the paradigm human society should emulate. According to both the old and the new evolutionary ethics, nature provides the standard by which we measure what is good and the goal toward which we should be striving. What has been determines what ought to be.

This position is rightly called the *naturalistic fallacy*. At the heart of the naturalistic fallacy is the question of how the good should be determined. "The idea of the good [*tou agathou idea*] is the highest knowledge," said Plato.[24] This high knowledge differs from ordinary knowledge of ordinary things. Knowledge of the good involves what is transcendent. As transcendent it is elusive. It lures us toward what is higher and, in doing so, forms or transforms our inner soul. The good is that toward which all things strive, and we are shaped by the very striving. It stands over against us, never fully realized, yet beckoning us to move toward it. The good is self-defining, not subject to our preferences or projections. It orients the human self, not the other way around. "Concerning the content of the Agathon nothing can be said at all," comments Eric Voegelin. "That is the fundamental insight of Platonic ethics. The transcendence of the Agathon makes immanent propositions concerning its content impossible."[25]

When in the early twentieth century philosopher G. E. Moore analyzed the naturalistic fallacy, he began with the central question of the field of Ethics: what is good? As we found with Plato, the good is somewhat elusive for Moore because it is "simple, unanalysable, indefinable."[26] This is decisive for discerning the naturalistic fallacy at its most abstract level. This fallacy is committed whenever one defines what is good "by reference to a *natural object* ... Those theories of Ethics, then, are 'naturalistic' which declare the sole good to consist in some

[23] Charles Sanders Pierce, *The Essential Pierce, 1867–1893*, Nathan Houser/Christian Kloesel, ed., (Bloomington: Indiana University Press, 1998) 357.

[24] Plato, *Republic*, VI: 505.

[25] Eric Voegelin, *Plato and Aristotle*, Volume III of *Order and History* (Baton Rouge LA: Louisiana State University Press, 1957) 112.

[26] George Edward Moore, *Principia Ethica* (Cambridge: Cambridge University Press, 1966) 37.

one property of things, which exists in time; and which do so because they suppose that 'good' itself can be defined by reference to such a property."[27] The fallacy is committed whenever one identifies "the simple notion which we mean by 'good' with some other notion."[28]

Certain constraints imposed upon us by nature result in aspects of our life that we deem necessary. But, we might ask: if something is necessary does this provide sufficient reason to deem it good? No, answers Moore. We might have to rely on what is necessary to devise a means toward a good end; but nature itself cannot decide what that good end ought to be. Just because something is natural does not make it good. "To argue that a thing is good *because* it is 'natural', or bad *because* it is 'unnatural', in these common senses of the term, is therefore certainly fallacious."[29] What is good may coincide with what is natural or it may stand in judgment over against what is natural; but the good is self-defining. It is not derivative from what is natural.

Moore applied his analysis of the naturalistic fallacy to the evolutionistic ethics of his day, the 1920s, which took the form of Social Darwinism. He identified the circular reasoning of associating what is "more evolved" with what is "higher" and therefore "better" because it is "more evolved." Survival of the fittest simply means the fittest survive; yet ethical sanction here is gained simply by surviving, by being more highly evolved. "Evolutionistic Ethics ... is the view that we ought to move in the direction of evolution simply *because* it is the direction of evolution. That the forces of Nature are working on that side is taken as a presumption that it is the right side ... It can only rest on a confused belief that somehow the good simply *means* the side on which Nature is working."[30] With such fallacious reasoning and egregious disregard of the intrinsic dimension of the good, Moore can only conclude, "Evolution has very little indeed to say to Ethics."[31] In short, ethical grounding requires appeal to what is transcendent; and the attempt to find an immanent grounding turns out to be fallacious.

[27] Ibid., 39, 41; italics in original.

[28] Ibid., 58.

[29] Ibid., 45; italics in original.

[30] Ibid., 56; italics in original.

[31] Ibid., 58. Theologian Douglas John Hall is reluctant to follow the road paved by liberal Protestantism that incorporated evolutionary progress into their theological scheme. "Nothing at all guarantees such an evolution of the Good. Nothing at all can be assumed about the future as improvement over the past." *The Cross in Our Context* (Minneapolis: Fortress Press, 2003) 223.

From Social Darwinism to Sociobiology

How does this play out on the stage set by sociobiology? E.O. Wilson lays down the gauntlet: he wants to base ethics on the empirical study of evolution and disassociate himself from any appeal to a transcendent good. In opposition to theologically or religiously based ethics, Wilson proffers a secular argument that moral guidelines are a product of the human mind that is biologically formed and regulated. "The empiricist argument, then, is that by exploring the biological roots of moral behavior, and explaining their material origins and biases, we should be able to refashion a wiser and more enduring ethical consensus than has gone before."[32] Wilson deliberately eschews religious foundations for ethical principles by denying the veracity of religious claims. Religious claims can be reduced to material explanations. "Ethical and religious beliefs are created from the bottom up, from people to their culture. They do not come from the top down, from God or other nonmaterial source to the people by way of culture."[33]

Key here is that he wants an immanent or materialistic ethic without reference to a good that transcends it. No revelation based source of the good will be allowed here. Wilson has also precluded appeal to the good as Plato would see it, namely as transcendent. And, as we will see shortly, he has precluded appeal to the good as Moore would see it, as self-defining and beyond analysis.

Where does Wilson go for his equivalent of what is good? He goes to what evolution has produced: "the universal bottom line of Darwinian genetic fitness: greater longevity and a secure, growing family."[34] Altruism or cooperative behavior has proven in community to foster longevity and growing families, so this must be incorporated into sociobiological ethics. "Cooperative individuals generally survive longer and leave more offspring."[35] These are the biologically established bases for moral guidelines. They are our Darwinian guide for future society.

Wilson bristles at the challenge of the naturalistic fallacy. He attacks the reasoning behind it. "The posing of the naturalistic fallacy is itself a fallacy," he writes. "For if *ought* is not *is*, what is? To translate *is* into *ought* makes sense if we attend to the objective meaning of ethical precepts ... From the consilient perspective of the natural sciences, they [ethical precepts] are no more than principles of the social contract

[32] Edward O. Wilson, *Consilience* (New York: Alfred A. Knopf, 1998) 240.
[33] Ibid., 247.
[34] Ibid., 252.
[35] Ibid., 253.

hardened into rules and dictates, the behavioral codes that members of a society fervently wish others to follow and are willing to accept themselves for the common good."[36]

On the surface, it appears that Wilson has scored a significant point: the *ought* itself has the status of an *is*. What we believe *ought* to mean in our society is itself derived from our evolutionary biology rising up via epigenetic rules into culture.[37] Philosophically, Wilson is saying that if values that project what ought to be are themselves existing within time and space like other things are, then any *ought* would have the status of what *is*. It would seem to follow, then, that Wilson's *is* – namely, evolution's principle of reproductive fitness – can legitimately serve as our *ought*. So goes Wilson's apparently forceful logic. Yet, below the surface, we can see that Wilson's argument fails for three reasons.

First, the character of the *ought* distinguishes it from any other thing that *is*. What characterizes something that ought to be the case is its inherent denial of dependence on what is the case already. Moral mandates draw us toward what ought to be in contradistinction to what currently is. This is what *ought* means.

Second, Wilson conflates the good with a property of nature. He conflates high knowledge with ordinary knowledge. As we saw in Plato and Moore, the good is not one more state of affairs set within our time. The good transcends all states of affairs by definition. The good is that toward which all things strive; and they are judged by it. In the last analysis, any evolutionary ethic can only say tautologically: keep on evolving!

Third, Wilson has falsely assumed that what we ought to do implies that we are able to do it. It is possible for us to have an ethical goal that we are unable to attain. The good is self-defining; it does not depend on what we are able or unable to do. Wilson's ethic merely says we should keep on doing what evolution has enabled or even programmed us to do all along. Mikael Stenmark criticizes Wilson here. "To argue [as Wilson does] in this way is not to reject the naturalistic fallacy. It

---

[36] Ibid., 249–250, italics in original. Richard Dawkins, curiously, begins with a genetic determinism and then announces that we human beings can leap morally beyond where nature has brought us. It is possible for us to advance to "disinterested altruism – something that has no place in nature." *The Selfish Gene* (Oxford: Oxford University Press, 1976, 1989) 201. Unlike Wilson, Dawkins wants to avoid committing the naturalistic fallacy. "'You can't get an ought from an is' has been said so often it is in danger of becoming tedious. It is none the less true for that." *The Ancestor's Tale: A Pilgrimage to the Dawn of Evolution* (Boston/New York: Houghton Mifflin Company, 2004) 211.

[37] "Epigenetic rules" in Wilson's vocabulary differs from the way I use "epigenetic" in this book. For Wilson, epigenetic rules mark constraints the genes place on culture. For me, epigenesis marks the emergence of new orders of complexity in the evolutionary process.

is rather to make use of a norm widely accepted among philosophers, namely that *ought implies can*. If a person ought to do a certain action then it must be possible for that person to perform this action. So, if our biology makes it impossible to treat other people in a fair way then we cannot blame ourselves for this."[38] Over against both Wilson and those philosophers whom Stenmark alludes to, we defend a notion of the good which derives from a transcendent source, God, and which will remain good whether we can or cannot realize it fully in the time in which we live.

Michael Ruse announces that he is willing to stand up and defend evolutionary ethics and take on all criticisms, even the naturalistic fallacy. He begins by observing historically that the *laissez faire* capitalism justified by Spencer's Social Darwinism fed the prior interests of the likes of John D. Rockefeller and Andrew Carnegie, who were already committed to *laissez faire* capitalism. "People took their prior prejudices – convictions, if you like – and read them into their evolutionism, and then read them right back out again as justified."[39] So, suggests Ruse, we ought not to blame Darwin for Social Darwinism. In fact, it would be better to call it "Social Spencerianism."[40]

From this historical observation Ruse moves to a sociobiological doctrine he shares with Wilson namely, that ethics along with other forms of culture is the genes way of making more genes. "Ethics is a collective illusion of the genes, put in place to make us good cooperators. Nothing more, but also nothing less."[41] In short, the selfish genes have given rise to ethics, which in turn support the replication of selfish genes. Ethics begins in nature and returns to nature. If ethics is so thoroughly natural then, are we to conclude, the naturalistic fallacy is avoided?

The Ruse argument like the Wilson argument falls short of persuading, because it does not take into account an important element in ethics, namely, ethics provides a moral vision of what ought to be, not what is. When it comes to Christian ethics, one might say that sociobiology could very well be describing what is, while contradicting what morally ought to be. The ethical vision is a transformative vision.

The good is a call to go beyond being, a lure to draw us toward the transformation of what is. For Plato the good is that which all people sense they lack and for which they strive. Although we may deceive

---

[38] Mikael Stenmark, *Scientism: Science, Ethics and Religion* (Aldershot UK: Ashgate, 2001) 69.

[39] Ruse, *Evolution and Ethics*, 34.

[40] Ibid., 30.

[41] Ibid., 47.

ourselves from time to time regarding just what the good means for us, the striving for what lies beyond is an inescapable part of the ethical *ought*. "The good is that which man does not yet possess conclusively, that which he must still strive to realize," writes Wolfhart Pannenberg; "The good thus contains in itself the difference between value and being. The good is 'beyond being', Plato said."[42]

Here we have applied Gilkey's hermeneutic of secular experience to the theory of evolution as employed by two forms of Social Darwinism, the evolutionary ethics of the early generation of Darwinists and today's sociobiologists. In each case, the attempt to ground an ethical program of moral *ought* on scientific observations in biology is fallacious. To project an ethical vision of what *ought* to be in continuity or especially in contrast to what *is* the case requires appeal to a transcendent apprehension of what the good is. Normally, such an ethical vision is constructed on one's belief in God and God's plan for the future. Attempts to employ science to muffle the voice of God go beyond the limits of what science legitimately should be doing.

Non-Adaptive Modern and Postmodern Theology: Huston Smith

The question we should pose to science and to the ideologies that frequently accompany science is this: where are the windows to transcendence? One answer is to shake the scientific dust off one's shoes, take leave of the modern world, and walk back to the home of premodern metaphysics where the windows are still open.

This is what Huston Smith advocates. Smith says that the metaphysics of modernity snuffs out the religious voice, which has become but an echo of our dismissible past. To live in the world of modern materialism is like living in a studio sound booth, where the padding prevents any transcendent noise from entering. We can only hear ourselves speaking into a microphone that amplifies what we say to a select audience. The ancient voice of transcendent reality is more than merely muffled. It has been totally shut out. At least, this is the way Huston Smith sees it.

I mentioned above that modernity forks into two postmodern directions, one holistic and one pluralistic. Huston Smith follows the holistic branch of this fork. In the process he decries the hegemony

---

[42] Wolfhart Pannenberg, *Theology and the Kingdom of God* (Louisiville: Westminster John Knox Press, 1969) 106.

of modern science and tries to fight back by retrieving the forgotten truths of premodern religion.

Few have held out with greater tenacity for opening our ears to the voice of premodern Christianity than the inveterate Huston Smith. This child of the University of Chicago way of treating the philosophy of religion has become the world's best known and most widely read interpreter of the world's major religious traditions. Smith calls himself a "comparativist," which reflects the era of scholarship in the immediate post World War II period. His comparisons of the different ways religious seers view reality is eclipsed, however, by what he finds in common. What he finds in common is a forgotten truth, a truth about the subtle spiritual dimension of reality that has been lost to consciousness by cultural Alzheimer's.

The religious worldview is important. It is important because it provides a purchase on reality. Deep within the human soul is a thirst, a thirst for reality. Mircea Eliade called this ontological thirst We human beings are born with a thirst for being, a drive toward ultimate reality.[43] Nothing less will quell the yearnings of our innermost soul. Huston Smith believes our ancient ancestors caught a glimpse of the ultimate; and they gave voice to what they glimpsed in a metaphysical view of the world that is worth retrieving.

Here is the structure of the world in which our predecessors lived as described by Smith and embellished by me. Let me list seven characteristics. First, there is an objective reality that speaks to us. Reality is not merely a social construction. We are a product of reality; we do not create it according to our social location or for personal therapy.

Second, reality is a mixture of the finite and the infinite. The infinite is awesome, and it includes the finite while transcending the finite.

Third, dualism is essential. Although this may appear to be a substance dualism of the Cartesian type, premodern dualism is more subtle. Ancient dualism begins with both the distinction and the continuity between the finite and the infinite, distinguishing them but not separating them. Our mundane reality is inextricably linked to transcendent reality, a truth which we fail to recognize when we focus solely on what is mundane, material, and finite.

Fourth, reality is organized hierarchically. A ladder of being links all levels, with the fulfillment of being on the top with near nothingness approaching the bottom. Spirit is higher than matter; eternity is higher than temporality; and infinity is higher than finitude.

---

[43] Mircea Eliade, *The Sacred and the Profane* (New York: Harcourt, Brace & World, 1957) 64.

Fifth, being and value come together. Each rung on the ladder of being is value-laden, with virtues transcending mere physical or material existence. Virtue is eternal, whereas corporal life is temporal and ephemeral.

Sixth, we find a dialectic between unity and diversity, between simplicity and complexity, between the absolute and the relative. At the highest level of being – the fullness of being – every separate level becomes unified in the Simple One, in the Absolute. The Absolute is perfect, and all that has being participates in this perfection. We might say that, in the last analysis, the ladder of being folds up into the Absolute and disappears into the simplicity of perfection.

Seventh, we finite human beings are unable to perceive and comprehend the ultimate perfection, the sublime Absolute. It must be revealed to us. We must receive its revelation. The key to unlocking the meaning of the revelation is symbolism, the religious language of symbol. Symbols involve doublespeak. They speak in concrete images while opening our consciousness to transcendent reality. What we learn literally opens windows to a mysterious dimension that lies beyond rational comprehension and possession. By peering into the shrouds of mystery we can dimly perceive the presence of perfection.

This worldview is shared by Christianity and certain non-Christian traditions alike. It constitutes the "universal grammar of religion," a grammar that comes to articulation in the symbolic speech of classical Christians. If we compress the "Christian worldview" into a sentence, Smith says: "the world is perfect, and the human opportunity is to see that and conform to that fact."[44]

When Huston Smith turns specifically to the question of evolutionary theory, he finds himself sympathetic to the Anthropic Principle with its fine tuning of the universe and even to the Intelligent Design movement.[45] Why? Because Intelligent Design theory opens evolutionary windows to transcendent reality. By observing advances in the design and complexity of organisms in evolutionary history, this theory proposes that only a transcendent designer – an intelligent designer – could be responsible. This dovetails with Smith's retrieval of the premodern worldview, where a second reality hides just behind the material reality studied by the natural sciences.

In sum, Huston Smith represents eloquently and forcefully a position that refuses to adapt to the modern mind, especially the mind dominated by the reductionist materialism that appears to accompany modern science. The religious vision in general, and Christian theol-

---

[44] Huston Smith, *The Soul of Christianity* (San Francisco: Harper, 2005) 33.
[45] Huston Smith, *Why Religion Matters* (San Francisco: Harper, 2001) 176–178.

ogy in particular, should not cow tow or pay obeisance to a worldview that has arbitrarily and unnecessarily closed the windows to its own transcendent depth and ultimate meaning. The Darwinian model of evolutionary biology could not, according to Smith's nonadaptive method, provide a worldview hospitable to a theology of divine action in the world.

Non-Adaptive Non-Theology: Intelligent Design

The Intelligent Design (ID) school of thought favored by Huston Smith connects the two realms, the mundane and the transcendent, with what they refer to as an "intelligent designer." The self-imposed restriction by modern scientists to limit all explanations of natural phenomena to this-worldly or naturalistic principles is unnecessarily constrictive, argues ID. The effects of transcendent interventions are present in the complex designs we observe in nature, yet evolutionary biologists operating within the Darwinian model are blind to the transcendent.

This self-imposed blindness on the part of Darwinian evolutionists is due to anti-religious ideological commitments, complains Philip Johnson. "Darwinism was not just a theory of biology, but the most important element in a religion of scientific naturalism."[46] Because established scientists want to use their science to promote a naturalistic replacement for classic Christianity, assails Johnson, they are less than fully honest when subjecting their own theories – such as the theory of evolution – to critical scrutiny and testing by the evidence. "Prejudice is a major problem ... Darwinism plays an indispensable ideological role in the war against fundamentalism. For that reason, the scientific organizations are devoted to protecting Darwinism rather than testing it."[47] Johnson intends for his arguments to operate within the sphere of science, on the plane of discourse where empirical evidence will support or challenge the dominant theory, the Darwinian model of evolution.

Others in the Intelligent Design camp attempt to supply empirical evidence from within the evolutionary picture of biology. Michael Behe, for example, advances the thesis that structures have arisen in biological history that are irreducibly complex – that is, they are designed. By "irreducible complexity" he means that the structure will not function if one part of it is removed. "By *irreducibly complex* I mean a single system composed of several well-matched, interacting

---

[46] Phillip E. Johnson, *Darwin on Trial* (Downer's Grove IL: InterVarsity Press, 1991) 150.

[47] Ibid., 153.

parts that contribute to the basic function, wherein the removal of any one of the parts causes the system to effectively cease functioning."[48] Such irreducibly complex designs could not have arisen in evolutionary history without the intervention of an intelligent designer, a designer who transcends the natural processes.

"Irreducible complexity" becomes "specified complexity" in the work of William Dembski. Dembski offers a method for determining if a structure is specifically complex. Looking at evolution through an explanatory filter, Dembski can ascertain whether or not specified complexity exists in a particular case. The filter poses three questions is the organism or event or object contingent? is it complex? and is it specified?

Specified complexity in evolutionary development cannot adequately be explained as arising by chance, argues Dembski. Dembski introduces a universal probability bound of 1 in 10^{150} as the limit below which it is improbable that chance was the source of the event or object. This number is a product of the estimated total number of elementary particles in the universe (10^{80}) times the age of the universe in seconds (10^{25}) times the number of events that can occur per second (10^{45}, or the Planck time).[49] Further, an event or object is said to be specified if the pattern detected is detachable from the event or object itself. Because some phenomena in nature yield "yes" answers to his explanatory filter, Dembski feels confident in concluding that such structures could not have arisen by gradual Darwinian step-by-step processes. Leaps in development must have occurred, and these leaps in complex design cannot be explained by random variation and natural selection. Complex design can be explained only by appeal to an intelligent designer who transcends the natural processes.

The Darwinists fight back. Richard Dawkins, who embodies the very ideology Johnson opposes, argues that Intelligent Design fails to make its case that complex design could not arise by chance. Dawkins says that natural selection provides an adequate explanation for gradual development; and natural selection is actually a mechanism that is much more specific in its direction than chance. No appeal to an intelligent designer is warranted, says Dawkins. Then he adds that ID "raises an even bigger problem than it solves: Who designed the designer?"[50]

---

[48] Michael Behe, *Darwin's Black Box: The Biochemical Challenge to Evolution* (New York, Touchstone/Simon Schuster, 1996) 69, Behe's italics.

[49] William Dembski, *No Free Lunch: Why Specified Complexity Cannot Be Purchased without Intelligence* (Lanham MD, Rowman & Littlefield, 2002) 22.

[50] Richard Dawkins, "Why Intelligent Design Fails," *Science and Theology News* 6:4 (December 2005) 9.

Robert John Russell makes an argument similar to Dawkins. The logic of the Intelligent Design argument cannot discriminate between "intelligent aliens" and what most people know as God. Because of the regress implied in the logic of the ID argument, the problem of the source of intelligent design is simply pushed back to the designer of the aliens, or the designer of the intelligent designer.[51] The link to the transcendent is suggested by not established.

Theistic evolutionist William R. Stoeger, S.J., at the Vatican Observatory is willing to accept as adequate the naturalist account of complex design in nature. "There is a deep order in nature, and it appears that the universe is fine-tuned for complexity and life – but uses a degree of chance and randomness within that larger framework of law and order to explore and realize the rich potentialities of that order."[52]

On one level, Intelligent Design appears to have adapted to the modern scientific worldview. ID arguments for acknowledging transcendent reality are offered at the scientific level, even if accompanied in some instances by acrimonious rhetoric. ID intends to make specifically scientific arguments and to be considered a scientific theory. On another level, the Intelligent Design position presupposes the premodern worldview as defended by Huston Smith.

What is valuable about both Smith and Intelligent Design is their valiant defense of belief in a transcendent God who is active in the mundane world. Yet, their demanded modifications of the scientific world picture may be premature. There are advantages in modern scientific methods for making knowledge of the natural world grow and expand and ready us for developing new technologies. The persistent pursuit of naturalistic explanations is what yields growth in knowledge, even if it is secular knowledge. The interjection of the idea of transcendence prematurely into explanatory schemes risks shutting off research too quickly.

Smith and Intelligent Design seem to make God one mundane cause among others within the course of natural processes. Might there be a way to avoid an unnecessary competition between divine causation and natural causation? This is what the theistic evolution school would propose.

---

[51] Robert John Russell, "Intelligent Design is Not Science and Does Not Qualify to be Taught in Public School Science Classes," *Theology and Science* 3:2 (July 2005) 131.

[52] Nandagopal R. Menon, "A Matter of Interpretation: An Interview with William R. Stoeger," *Science and Theology News* 6:4 (December 2005) 26.

Adaptive Modern and Postmodern Theology: Theistic Evolution

Other theologians find that we must adapt to modernity, especially to the view of reality spawned by evolutionary theory. Jesuit paleontologist and evolutionary theist Pierre Teilhard de Chardin finds it inconceivable that we would bypass the Darwinian scheme. Yet, he recognizes the difficulties. "What makes the world in which we live specifically modern is our discovery in it and around it of evolution. And I can now add that what disconcerts the modern world at its very roots is not being sure, and not seeing how it ever could be sure, that there is an outcome – *a suitable outcome* – to that evolution."[53] What Teilhard adds to the uncertainty of the science is something only theology can provide, namely, confidence in a glorious outcome due to a promise given us by God. That outcome is suitably named by Teilhard, *Point Omega*.

Princeton's Wentzel van Huyssteen greets the advent of Darwinian theory with welcoming applause. He draws Darwinian epistemology right into the home of religious knowledge. He invites the concept of evolution to walk right into the living room of Christian theology, to transform theology into a modern and even postmodern apprehension of reality. "Our theological reflection is radically shaped not only by its social, historical context, but also fundamentally shaped by the biological roots of human rationality. In contemporary evolutionary epistemology, then, we will find surprising, if not startling, attempts to facilitate precisely the challenge of a constructive form of postmodernism: the need for a more comprehensive and integrated approach to the problem of human knowledge."[54]

Teilhard and van Huyssteen have company in a larger school of thought, *Theistic Evolution*. Those who adhere to theistic evolution did not name themselves. The label was coined by the Scientific Creationists as a term of derision. Even so, it seems an apt enough label for a group of theologians who find their fundamental understanding of God's action in the world compatible with evolutionary theory.

In a book I co-authored with my colleague, Martinez Hewlett, we located the theistic evolutionists on a spectrum of divine action. Ranging from interventionist theism on one end to deism and atheism on the other, some differences in method for establishing compatibility

---

[53] Pierre Teilhard de Chardin, *The Phenomenon of Man* (New York: Harper, 1959) 228.

[54] Wentzel van Huyssteen, *Duet or Duel? Theology and Science in a Postmodern World* (Harrisburg PA: Trinity Press International, 1998) 134–135.

between theology and evolution can be discerned. The creationists, especially the Scientific Creationists, hold that God creates the world in the beginning (*creatio ex nihilo*); and at this original point God fixes the species. No evolution from one species to another takes place. Next to the creationists we place Intelligent Design, which locates divine action within the speciation process; here God the intelligent designer intervenes in evolution to precipitate leaps forward in levels of complexity.

At the other end of the spectrum we place ontological materialists who utilize Darwinian theory to justify agnosticism and atheism. No god exists, according to this view. Next to the agnostics and atheists we place the deists, who hold like the creationists that God established the creation at the beginning; but now, according to the deists, God has withdrawn to allow nature to evolve completely according to its own natural laws.

The theistic evolutionists fit right in the middle. On the one hand, they affirm God's action in the natural world. On the other hand, they affirm self-organization within nature; and they accept the history of speciation explainable by random variation and natural selection. Most theistic evolutionists hold to a noninterventionist interpretation of divine action – that is, God is involved in the evolutionary process but not in a manner that upsets or disrupts. B.B. Warfield, for example, posited a *concursus* or concurrence of God's action and nature's action in creating the human race through evolution.[55] Robert John Russell, to cite another example, posits a noninterventionist objective divine activity at the quantum level to account for random variation in mutations.[56] Such approaches make God's relationship to evolutionary biology non-miraculous, one might say.[57]

Theistic evolutionists have begun with the Christian faith first formulated in an ancient culture now long out of date, and they have sought to update it by rendering it compatible with the world picture painted by evolutionary science. One of the toughest questions each theistic evolutionist must pose has to do with the role of divine action in nature. Some posit an interventionist God, others try to formulate a theory of divine action without intervention. In all cases we see theology as faith seeking expanded understanding of our world. However,

---

[55] B.B. Warfield, *Evolution, Scripture, and Science,* ed. by Mark A. Noll/Daniel N. Livingstone (Grand Rapids MI: Baker Books, 2000).

[56] Robert John Russell, "Special Providence and Genetic Mutation: A New Defense of Theistic Evolution," in *Evolutionary and Molecular Biology*, ed. by Robert John Russell/William R. Stoeger, S.J./ Francisco J. Ayala (Vatican City State/Berkeley: Vatican Observatory and CTNS, 1998) 191–224.

[57] Ted Peters/Martinez Hewlett, *Evolution from Creation to New Creation* (Nashville: Abingdon, 2003) 31.

we will not follow the implications of this observation any further at this point. We will change our direction slightly, back to the question of hermeneutical windows to transcendence.

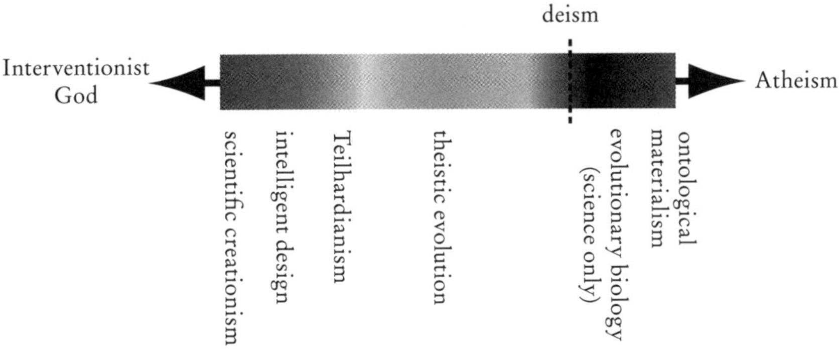

Figure 2: Divine Action by Martinez Hewlett. Used with permission.

The Hermeneutics of Demythologizing

One item to notice about all of the positions engaging evolutionary theory is they make the same assumption, namely, what we say about God and what we say about nature belong on the same epistemological plane. The kind of knowledge we posit about God is identical with, or at minimum overlaps with, the kind of knowledge we gain from the empirical sciences. This assumption marks a remarkable shift from the way hermeneutics was pursued at the middle of the twentieth century, at least in Europe.

Rudolph Bultmann defined the task of hermeneutics in such a way that we could distinguish between the worldview or world picture, on the one hand, from the existential message of the Christian gospel, on the other. The gospel was first formulated within a biblical world picture that is now out of date. It has been replaced by modern science. Yet, we can distinguish the naked gospel from the clothes in which it comes dressed.

That ancient worldview was a mythological worldview, that included supernatural interventions into natural events. Such supernaturalism is no longer admitted to the scientific picture. Matching the biblical worldview with the scientific worldview is neither possible nor desirable. What Huston Smith proposes as a retrieval of the premodern world picture would be impossible for Bultmann. "It is impossible to repristinate a past world picture by sheer resolve, especially a *mythical* world picture, now that all of our thinking is irrevocably formed by science."[58]

What shall we do? Bultmann's answer is to demythologize the gospel, to disentangle the essential Christian message of God's grace in Jesus Christ from the premodern worldview. In fact, says Bultmann, the essential message or *kerygma* should be disentangled from all world pictures, including the modern scientific picture. To substitute the scientific world picture for the mythological one would perpetuate a confusion that our relationship to God depends on our understanding of the world. God is other-worldly, so we need to disconnect our self-understanding of existing in relationship to God from our mundane understanding of the way the world works.

We need to ask ourselves whether we want to follow Bultmann or not. It is my judgment that Bultmann is right epistemologically that our relationship to God is not reducible to our world picture, whether mythical or scientific. Our relationship to God is not reducible to any concepts or ideas; rather, it is personal. Our relational connection to God exists within the depth of our souls at a prearticulate level, where only the Holy Spirit can speak for us. Romans 8:26 "The Spirit helps us in our weakness; for we do not know how to pray as we ought, but that very Spirit intercedes with sighs too deep for words."

Having said this, does this imply that our faith relationship to God has no relationship at all to our world picture? I do not recommend we try to maintain a total divorce. Rather, I recommend we try to construct a world picture that takes the intimacy of our relationship to God and to God's promises as the foundation. Such a world picture will be a speculation, to be sure; yet, speculate we must. This will be a speculation based upon the insight of faith, not a prelude to faith.

For Saint Anselm faith seeks understanding, *fides quaerens intellectum*.[59] For Saint Thomas Aquinas theological science is a form of wisdom because it orients all things in the created order to God, "the highest cause in that order."[60] It seems to me that the task of the theologian in our era is to seek to understand the implications of the Christian faith in light of a world picture painted by modern science. And, whether we like it or not, our world picture is dominated by icons of evolution. Just how should faith rise to understanding in an evolutionary view of the universe?

---

[58] Rudolph Bultmann, *New Testament and Mythology: And Other Basic Writings*, ed. and tr. by Schubert M. Ogden (Minneapolis: Fortress Press, 1989) 3.

[59] Anselm, *Proslogion: Fides Quaerens Intellectum.*

[60] Thomas Aquinas, *Summa Theologica*, I.1:Q.1:Art.6.

Just to clarify, the method I am proposing here differs from classical understandings of revelation. In the premodern period, Christian theologians believed that their metaphysics were revealed, the result of a theophany. In the postmodern period, I suggest theologians view metaphysics as a speculative discipline. Building on a prearticulate experience of Jesus Christ in faith, the theologian constructs a world picture where all things relate to the God of faith.[61] I advocate a postcritical method of worldview construction through which the Christian faith seeks to understand itself along with all of reality.

The enduring contribution of Bultmann's hermeneutics is this: what is essential to the kerygma is not reducible to what is signified on the plane of scientific discourse, or any other worldview discourse for that matter. Yet, the theistic evolutionists must be applauded for seeking to understand our evolutionary world in light of God, the first and highest cause. What this leads to is a commitment to find consonance where possible between the Christian view of God as creator and redeemer, on the one hand, and the elements of evolutionary biology with its deep time and speciation, on the other.

Enough said about this at this point. What we need to learn from Rudolph Bultmann as well as Huston Smith is the following: this world has windows open to transcendent reality. Furthermore, if our world picture closes those windows, it is premature and unnecessary. One of the formidable barriers to a healthy religious understanding of reality is reductionist materialism which, though not identical to science, so often accompanies science. Both Smith and Bultmann in their different ways provide reasons for saying "no" to a world picture that arbitrarily excludes the divine. How to say "no" exactly where necessary and "yes" where that is necessary provides the challenge taken up by Theistic Evolutionists.

A Theistic Evolutionary Proposal

John Polkinghorne asks just what theologians should do when confronting modern science. Two mistakes have been made since the Enlightenment. One is to place theology on the same plane of explanation as science. The other is to try to reinject supernaturalism into science by

---

[61] I dub this method of constructive theology "postmodern" because I follow Alfred North Whitehead's post-Cartesian understanding of metaphysics as speculative. *Process and Reality: Corrected Edition,* ed. by David Ray Griffin/Donald W. Sherburne (London/New York: Free Press, 1929, 1978) 3. Similarly, I follow Paul Ricoeur's post-critical rational construction of pre-critical meaning at the symbolic level of experience. *The Symbolism of Evil* (Boston: Beacon Press, 1967) 347–356.

claiming that God began it all. "One can see that two theological mistakes had been made. One was the attempt to rival science on the latter's own explanatory ground ... The second error was to take too narrow a view of God's means of creation, as if discontinuous decree was to be preferred to the stealthy continuity of the outworking of the divinely ordained laws of nature. *Continuous* creation is a proper concept for the theologian to consider."[62] In what follows here, we will take a look at a "stealthy" understanding of God's continuous creation.

Together with my colleague in molecular biology and virology, Martinez Hewlett, we have begun to construct a world picture rooted in the gospel kerygma with branches informed by evolutionary biology. These roots are dug deep in the transcendent, so naturalistic explanations will limit the amount of wavering. Here in brief outline is our speculative proposal which, we hope, expands the understanding of our faith.

1. *God creates over deep time, not only once at the beginning.* God's creative work is not yet complete. It is still happening, and evolutionary history finds its place within God's ongoing creative activity. God is constantly opening up each moment in natural history with potential for developing new life forms, new things. What actually happens in nature is contingent, and God makes this contingency possible so that creation might progress.

With this in mind, we want to say that God creates from the future, not the past. God starts with redemption and then draws all of creation toward it. Or, perhaps better said, God's ongoing creative work is also God's redeeming work. Only a redeemed creation will be worthy of the stamp of approval we read in Genesis, "very good."

Even while affirming a historic moment of creation out of nothing, *creatio ex nihilo*, we also affirm as a complement continuing creation, *creatio continua*. Where we find ourselves now as heirs to our evolutionary ancestors is looking forward to God's omega, God's eschatological completion of the divine work of creating our world.

2. *We understand why evolutionary biologists do not perceive purpose within nature, because God has a purpose for nature that is not clearly visible within nature.* We recall Saint Paul saying that the power of God as creator is manifest in the creation, though it is somewhat difficult to see it unmistakably. To perceive the way nature is currently designed does not give us reason to think we understand the mind of God.

---

[62] John Polkinghorne, *Faith, Science, and Understanding* (New Haven CT: Yale University Press, 2000) 67–68.

It is our conjecture that God's overall purpose for the creation is not discernable within the creation. God's uniting purpose for the history of evolution is not discernable within biological structures or processes. This means an inner *telos* or purpose or design does not stand up and advertise itself. Therefore, we will not attempt to locate purpose or direction or even value *within nature*.

Instead, as Christians, we affirm a divine purpose *for nature*. We plan to look for this divine purpose where it belongs, namely, in God. The purpose for the long history of nature over deep time is not a built in design or direction.[63] Rather we prophetically contend that it will be retroactively imparted by God in the eschatological new creation.

Evolution is a form of history, biological history. Astute theologians have learned from the prophetic tradition to avoid immanentizing God's transcendent purposes; because the claim to possess the divine *telos* in one's own institutions leads to the demonic misuse of power. Philosopher of history Eric Voegelin issues the warning for history that we apply to evolution: "The attempt at constructing an eidos of history will lead into the fallacious immanentization of the Christian eschaton."[64] Rather than sticking God's eschatological *telos* within evolutionary history, we prefer to leave it transcendent to nature's history.

Therefore, we do not find such a purpose at creation's beginning (or during saltations of increased complexity, for that matter). Rather, we find it at the end, looking backwards. Purpose comes from what is final looking backward, not from a potentiality lying in wait at the beginning. In fact, the Greek word for end, *telos*, means end both as final state and as purpose or goal. God has a telos for nature, even if we can't see it within nature. It is the future act of redemption that determines what previous creation will have meant, and this can be discerned only eschatologically. It is omega that determines alpha.

3. *God is the primary cause, while the self-organization of life forms through evolution operates in the domain of secondary causation.* By God imparting the quality of openness to the future, God makes room for the distinction between primary and secondary causality. God's direct act is the primary cause. God establishes the world. God gives the world being, and preserves if from falling into nonbeing. God also

---

[63] Douglas John Hall concludes: "what the theology of the cross calls for is an eschatology that affirms both the transcendent purposing of the world, *yet without bypassing the* world, and the belief that purpose inheres in the structure of the world, *yet without assuming necessity.*" *Cross in Our Context*, 222.

[64] Eric Voegelin, *The New Science of Politics*, in *The Collected Works of Eric Voegelin* (34 Volumes: Baton Rouge/London: Louisiana State University Press, 1990–1995) 5:187.

imparts openness toward a future that can be different from the past. This permits the creaturely world to take action. This permits what we call secondary causation within evolutionary history, leading to unpredictable patterns of variation and self-organization.

As primary cause, God provides the world with a future in two sense. The first sense, as we have just said, is that God provides openness so that physical reality at each moment can reorganize itself and develop something new. The second sense of God's future is fulfillment. God gave the world a promise that, in the end, everything would be "very good." God provides the final cause, so to speak, at least in a qualified sense. Anticipating fulfillment, we want to say that future-giving is the way God both creates and redeems the world.

The distinction between primary and secondary causation makes it possible and legitimate for scientists to seek strictly natural explanations for natural phenomena. This does not mean that the scientific picture is the entire picture, however. God's primary causative action is always co-present, even if not accounted for.

Whether visible or invisible, each moment God imparts openness to the future that releases the present from bondage to past determinations. God's creative activity is never ceasing. Each moment the entire physical universe is given its existence in such a way that it is open toward what comes next. This ceaseless future giving by God explains why the laws of nature cannot grip nature in a rigid determinism. It explains why each moment has the freedom to transcend the previous moment. What we see as contingency or chance or self-organization is the result of God's liberating gift of an open future. We call this continuing creation, *creatio continua*.

4. *Genuine chance, contingency, and human freedom belong to the secondary processes that make up ongoing creation.* The knottiest challenge of the Darwinian model of evolutionary biology is chance. Chance and unpredictability belong to both random genetic mutation and natural selection. As we just said, we affirm both openness and purpose, but the purpose comes from God's future. It's not built in. It's imparted.

How do we combine these? We have two thoughts we need to hold together. On the one hand, God's gift of futurity to the physical world makes room for evolution by providing openness to change and self-organization. On the other hand, God's eschatological future embodies an aim, namely, the harmony and benefit of all God's creatures. Neither alpha nor omega belong to chance, even if much of what happens in between does.

In the meantime, we take advantage of holistic thinking, one branch of postmodern thinking. According to holistic or emergentist thinking, the whole is greater than the sum of the parts. Evolutionary history has witnessed the emergence of living creatures, which as organisms constitute wholes that reorganize and give new meaning to the chemical parts that make them up. No organism can be reduced to its chemical components and retain its identity as a living creature.

We observe that new wholes transform past parts. Integration into new, more comprehensive unities preserve while renewing what came before. This holistic complexification process is nonlinear. Adding a new whole changes an entire situation in a significant way. The degree of transformative effect renders redemption possible. Can we learn something theologically here? By analogy, might we apply what we have observed as emergent holism to God's eschatological promises? We answer affirmatively, even if only tentatively.

Conclusion

We opened with the hermeneutical question, asking just how theologians can interpret the Christian faith – originally coming to articulation in an ancient culture with an outdated mythological world picture – in today's cultural context dominated as it is by modern science, especially Darwinian evolutionary biology. I pointed out that one of the disconcerting aspects of the modern scientific context is that windows to transcendence seem closed. When science is done right, its methodological commitments lead to strictly this-worldly or naturalistic explanations; and when science is contaminated by ideologies such as ontological materialism or Social Darwinism it deliberately supports atheism. Even if theologians are willing to demythologize and extricate the biblical kerygma from its original worldview, theologians cannot endorse a world picture that is closed to transcendence. This calls for apologetic analysis as well as accommodating construction.

Finally, our context calls for some brand of theistic evolution. Exactly how we should construct a picture of our world where theology and evolution are compatible is less than crystal clear. That we should engage in such constructive work is demanded. In partnership with my scientific colleague, Martinez Hewlett, I have proposed a version of theistic evolution that begins with God's promise of new creation and then asks: just how do we get from here to there? The value of evolutionary biology is that it provides a model for watching our movement. It needs completion by theology, however, so we know what direction it is ultimately moving.

Chapter 6

The Post-Human Soul, Part One

Can we storm the gates of Omega with science and technology like the builders of the Tower of Babel sought to storm the gates to heaven? If not, should we project a utopian vision akin to creating our own Omega; then should we use this vision to guide us in that direction, accepting the fact that we will not arrive at the final destination on our own power alone? Do science and technology work with just such an implicit vision?

Science and technology as cultural phenomena are inextricably tied to promise, progress, and perfection. Science is the engine that drives progress, and technology steers it toward perfection. Like the cavalry riding in to rescue the settlers from otherwise certain demise at the hands of raiders, science is invested with saving us, especially saving us from physical imperfections. Science, especially in the form of medical science, can lead to dramatic improvements in the human situation. But, we might ask: is it realistic to ask medical science to deliver human perfection? Would human perfection include cybernetic immortality, a form of the disembodied soul lodged in a post-human computer of the future? Whether realistic or not, should the striving for perfection provide an ethical vision or guide?[1]

What counts as perfection? For John Wesley, Christian perfection is possible in this life through the work of the Holy Spirit within a loving heart. Can medical science replace the Holy Spirit? Wesleyan Methodists and Christians more generally aim at cultivating a loving and caring disposition: "pure love reigning alone in the heart and life, this is the whole of scriptural perfection."[2] What would count as per-

---

[1] This chapter draws upon research for another published piece, Ted Peters, "Perfect Humans or Trans-Humans?," for St. Deiniol's Library research project, *Future Perfect? God, Medicine, and Human Identity*, ed. by Celia Deane-Drummond. My background study of genetic therapy and enhancement has been pursued as part of a larger research project, "Altering Nature," at Rice University on a grant from the Ford Foundation with Andrew Lustig as Principal Investigator. The team researching enhancement issues includes Karen Lebacqz, Crawford Cromwell, Bernard Lo, and Estuardo Aguillar along with me serving as chair, 2005.

[2] John Wesley, *A Plain Account of Christian Perfection* (London: Epworth, 1952) 52.

fection for research science and perhaps medical technology? Would the perfection toward which medical therapy aims be a state of good health? Would the perfection toward which genetic or other physical enhancement aims be excellence, or superiority? Do such aims lead us so far beyond present capacities that we must talk about a trans-human or post-human future?

What role does a vision of perfection play? In Christian perfection, the vision of a loving heart draws the person of faith toward increased caring, toward transformation from selfishness to selflessness. In medical therapy, the vision of good health directs the steps to be taken by doctor and patient. In considering enhancement, the vision or model of the excellent or superior person suggests genetic or other biological alterations that could produce this effect. In assessing the emerging concept of trans-humanism, visions of good health and enhanced capacities carry our imaginations beyond what we previously thought were within the range of finite human existence.

With these questions and prospects in mind, I would like to examine carefully three forms of visioning that lead to transformation through medical science: therapy, enhancement, and transhumanism. All depend upon an image of who we are and who we would like to become. Enhancement and transhumanism each add an element of the fabulous, an element of excellence or superiority that takes us beyond good health. Do they require an implicit image of perfection? Or, is the mere striving for partial betterment sufficient to energize and direct the relevant medical sciences?

The track we will follow will be that of genetic or other biological interventions as they pass through therapy, enhancement, and transhuman projections of disembodied immortality. When we switch to the ethical and theological tracks, we find little or no controversy surrounding therapy; mild caution regarding enhancement; and rejection of at least the extreme forms of transhumanism.

One of the problems associated with enhancement and transhumanism is that they risk promising too much. They promise to make us into fabulous human beings, in the latter case even something post-human. They promise transformation, and transhumanism even promises transformation leading to a new level of perfection. Tacitly, they risk assigning science the job of providing the equivalent of salvation. Our risk is that our generation might be tempted to ask of science what only God can deliver. God may be able to deliver perfection. Science cannot.

The God of the Bible is a God of transformation, to be sure. The Easter resurrection of Jesus announces God's promise of a radical transformation yet to come: the eschatological new creation. In anticipation of the new creation, Jesus performed miracles of healing. We might call his healing "eschatological therapy." Those healed experienced transformation as today's effect of tomorrow's renewal. Built right into the Christian world view is the vision of a coming new creation accompanied by a ministry of healing in today's broken world. Any Christian ethic should embrace transformation aimed at healing, to be sure.

As we move from eschatological vision to bioethics, we need to distinguish between promises peculiar to God and those we can realistically assign to medical science. Ultimate transformation and perfection belong to God's agenda. Striving for well being and even flourishing belong on the human agenda, and this includes advancing scientific research as a means to that end. Knowing the difference between a vision of perfection and our appropriate striving will require wisdom and prudence.

Wisdom and prudence are what Celia Deane-Drummond offers. Her virtue approach to ethics is "eschatological in orientation; it explores through a particular theological telos, what might be the good end for humanity."[3] With the ultimate end envisioned, she then commits herself to honoring wisdom, which is "an expression of the eternal mind of God, while at the same time affirming what can be known in creaturely existence."[4] From wisdom emerges her virtue ethic. She follows Thomas Aquinas by placing prudence in special relation to the other cardinal virtues,: justice, fortitude, and temperance. Prudence is the application of wisdom to human affairs; and ethics counts as such a human affair. "The ability of prudence to be still, to deliberate well, is a quality desperately needed in the frenzied search for new methods and techniques in biological science that are considered to have particular usefulness for humanity."[5]

My proposal is that Christian deliberation over speculations regarding the next advances in neurocognitive research affirm at the level of assumption that transformation on behalf of improved human flourishing is attuned to God's eschatological promise; and, in addition, I propose that we invoke wisdom and prudence in assessing what is realistic about the promises of medical advance. I will conclude that therapy is an uncontroversial divine mandate, that transhumanism is unrealistic, and that enhancement deserves continued ethical monitoring and sorting out.

---

[3] Celia E. Deane-Drummond, ed., *Brave New World? Theology, Ethics, and the Human Genome* (London/New York: T&T Clark International, 2003) 235.

[4] Celia E. Deane-Drummond, *The Ethics of Nature* (Oxford: Blackwell, 2004) 20.

[5] Drummond, *Ethics of Nature*, 14.

From Therapy to Enhancement

For the sake of this discussion, I would like to distinguish three terms, *therapy, enhancement,* and *transhumanism.* The first two are commonplace in medicine and biotethics. By 'therapy' we mean healing, the addressing of a pathology for purposes of restoring health. By 'enhancement' we refer to medical measures that improve an individual's functioning or improve the human species beyond what had previously been thought to be its norm. Therapy is a response to a pathology, whereas enhancement initiates an improvement without reference to a pathology.

Therapy restores health. What is meant by health? The World Health Organization holds that "health is a state of complete physical, mental, and social well being and not merely the absence of disease or infirmity."[6] The late Pope John Paul II offers a similar definition. "From a Christian perspective, then, health envisions optimal functioning of the human person to meet physiological, psychological, social, and spiritual needs in an integrated manner."[7] Although such definitions may open themselves to ambiguity, the end or goal of therapy or cure gains some helpful specificity.

Even though good health may be pursued with passionate hopes for a cure, and even though therapy's achievement results in joy, good health is not a form of enhancement. To be healthy, we assume, is to exist at the norm, to live as our bodies were originally designed to live. Despite the joy good health brings to the healthy person, we would want everyone in our neighborhood to benefit from the same level of wellbeing.

Enhancement is distinguished from therapy here because it involves efforts to make someone not just healthy, but better than healthy. More than offering a cure, enhancement optimizes attributes or capabilities beyond what good health requires. The goal might even be to raise an individual from standard up to peak levels of performance. If therapy

---

[6] http://policy.who.int/cgibin/om_isapi.dll?hitsperheading=on&infobase=basicdoc&jump=Relations%20with%20NGOs&softpage=Document42#JUMPDEST_Relations%20with%20NGOs

[7] Pope John Paul II, "The Ethics of Genetic Manipulation," *Origins*, 13: 23 (1983) 385. We can distinguish two tendencies in current medical philosophy, one using a strictly biological model, which tries to define health and disease scientifically – that is, objectively and value free – and another, the holistic model, which takes all aspects of human being into account. In the holistic model the evaluative element is explicit. Even in the biological model normativity must be established, making evaluation implicit. "Discussion of disease and health is never merely a scientific matter, but always includes a value judgment as well," concludes Kris Dierickx, "Is Nature Neutral: The Concept of Health," in *Is Nature Ever Evil?* ed. by Willem B. Drees (London/New York: Routledge, 2003) 187.

via genetic intervention would bring an individual up to what is average or normal, then enhancement would bring an individual beyond the average or norm up to a level of excellence above others. Eric Juengst defines enhancement this way: "The term *enhancement* is usually used in bioethics to characterize interventions designed to improve human form or functioning beyond what is necessary to sustain or restore good health."[8]

The U.S. President's Council on Bioethics (COB) defines *human gene therapy* as directed genetic change of human somatic cells to treat a pathological situation, a genetic disease or defect.[9] By *human genetic enhancement* the COB refers to the use of genetic knowledge and technology to bring about improvements in the capacities of living persons or future generations.

From Enhancement to Transhumanism

Beyond enhancement, the open arms of transhumanism seem to be welcoming us. "Transhumanism is the view that humans should (or should be permitted to) use technology to remake human nature," write Heidi Campbell and Mark Walker.[10] Through genetic technology, information technology, and nanotechnology transhumanists believe the possibility exists for us to greatly enhance the healthy life span of persons, increase intelligence, and make ourselves happier and more virtuous.[11] The key is to recontextualize humanity in terms of technology. This leads to a vision of a posthuman future characterized by a merging of humanity with technology as the next stage of our hu-

[8] E.T. Juengst, "What Does *Enhancement* Mean?" in *Enhancing Human Traits: Ethical and Social Implications,* ed. by Erik Parens (Washington DC: Georgetown University Press, 1998) 29.

[9] "Staff Background Paper: Human Genetic Enhancement," U.S. President's Council on Bioethics (December 2002) http://www.bioethics.gov/. The term "enhancement" as currently used in genetics was introduced by W. French Anderson, "Human Gene Therapy: Scientific and Ethical Considerations," *The Journal of Medicine and Philosophy* 10 (1985) 275–291.

[10] Heidi Campbell/Mark Walker, "Religion and Transhumanism: Introducing a Conversation," *Journal of Evolution and Technology*, 14:2 (August 2005) 1. See: Nick Bostrom 2005, http://www.nickbostrom.com/tra/values.html.

[11] Progress in transhumanism will most likely depend on progress in nanotechnology, the manipulation of matter on the scale of atoms and molecules. A *nanometer* or *nm* (based on the Greek word *nanos* meaning dwarf), is one billionth of a meter. One nanometer equals the length of ten atoms of hydrogen side-by-side. Worldwide, government and industry invested $10 billion in nanotechnology R&D in 2004. See: Hope Shand/Kathy Jo Wetter, "Shrinking Science: An Introduction to nanotechnology," in *State of the World*, ed. by Linda Starke (New York: W.W. Norton, 2006) 78–95.

man evolution. *Posthuman* refers to who we might become if transhuman efforts achieve their goals. We are on the brink of becoming more than human, say the transhumanists.

"It seems to me," writes Estonian theologian Anne Kull, "that in the cyborg figure a symbol has been found to interpret our time and our technologies and ourselves. Our technical things have acquired uncanny liveliness of their own."[12] This symbolic role will be vividly illustrated as we turn to the connection between transhumanism and immortality.

For transhumanists, death should become voluntary. Immortality should become an option. Once we advance to the posthuman future, we will no longer be required to die. Immortality will belong to our definition as posthumans. Now, how do we get from here to there?

One step toward taking us from our present mortal human state to a posthuman immortal state is to build upon a key assumption, namely, that who we are as persons is centered in our brain activity. Transhumanism assumes that intelligence as a pattern of brain activity is the home of our personhood. It is the cognitive code or information pattern that is definitive. We are what we think, so to speak. Therefore, to improve the human situation we may improve on biological function insofar is it enhances cognitive function; and, if necessary, we might even transfer cognitive function into a machine. The goal of transhumanism is to download the contents of human consciousness onto a vast computer network and, via the network, achieve a kind of disembodied yet intelligent immortality. As a machine, we could enjoy perfections that the limits of biology prevent us from attaining.

Anne Foerst explains. She founded and directed the "God and Computers Project" at MIT's Artificial Intelligence Laboratory in the mid 1990s. She worked in the field of robotics, with AI entities known as "Cog" and "Kismet." Trained in both theology and computer science, she found her MIT colleagues believing "that humans are nothing but meat machines that carry a computer in their head. As soon as we have decoded the program that runs on the wetware of the brain, we can download it into the hardware of a computer and live forever. This is the major belief of a movement that calls itself *transhumanism* or *extropianism*."[13]

The postulated sequence goes something like this. First, AI researchers will simulate human intelligence in a computer, in a robot. Second, humans and machines will merge step by step, replacing portions of

---

[12] Anne Kull, "Exploring Technonature with Cyborgs," *Is Nature Ever Evil?* ed. by Willem B. Drees (London/New York: Routledge, 2003) 241.

[13] Anne Foerst, *God in the Machine: What Robots Teach Us About Humanity and God* (New York: Dutton, 2004) 43.

our brains with mechanical parts. Third, AI researchers will reduce existing human intelligence to a pattern of information processing and download this into a computer or a robot. This will constitute an evolutionary advance, actually a leap forward that could lead to cybernetic immortality – that is, immortal intelligent life in a machine that gets constant backups.

Ray Kurzweil predicts that by the end of the 21st century human beings will attain cybernetic immortality. Up until now, he says, our mortality has been tied to the longevity of our hardware. So, when our hardware crashes, our thought processes crash with it. When we instantiate ourselves in our computational technology, our software and hence our immortality will no longer be dependent on the survival of our body. Our immortality will be contingent on our being careful to make frequent backups.[14]

How likely is this to come to pass? Noreen Herzfeld, a computer scientist and theologian at St. John's University in Collegeville, Minnesota, has tracked earlier projections of AI achievements since the 1950s. The record is dismally poor. Goals have not been reached even now in the early 21st century, despite enormous progress in computer development. No computer yet in existence can be deemed intelligent. "While computing in general has advanced dramatically in the last fifty years, advances in AI have been limited. Neural networks remain at a level far below the complexity of the human brain. Current research in neuroscience suggests that the workings of the brain are far more complicated than was initially supposed and may not be capturable in neural net technology as we currently conceive it."[15]

In all three – therapy, enhancement, and transhumanism – we employ scientific research and medical technology to intervene in our natural processes to attain a certain end. The end for the first two is improved biological functioning. The end of the third is improved cognitive functioning with or without a biology.

---

[14] Ray Kurzweil, *The Age of Spiritual Machines* (New York: Viking, 1999), chapter 6. Ian Barbour is slow to grant the assumption that we can transfer human consciousness to silicon. "I suspect that it will turn out that conscious awareness requires forms of organized complexity or properties of neural cells and networks that have no parallels in silicon-based systems. I do not think we can exclude the possibility of conscious computers on metaphysical grounds, but there may be empirical grounds for the impossibility of computer consciousness ... I am willing to leave this question open." Ian G. Barbour, Neuroscience, Artificial Intelligence, and Human Nature," in *Neuroscience and the Person: Scientific Perspectives on Divine Action*, ed. by Robert John Russel/ Nancey Murphy/Theo C. Meyering/Michael A. Arbib (Vatican City State/Berkeley: Vatican Observatory and CTNS, 1999) 266.

[15] Noreen L. Herzfeld, *In Our Image: Artificial Intelligence and the Human Spirit* (Minneapolis: Fortress Press, 2002) 72–73.

Health and Enhancement: A Blurry Line?

When we compare the first two, we can see that the line between good health and enhancement might be difficult to draw, especially if both seek to optimize functioning in an integrated manner. The distinction is blurry, not sharp. According to the World Council of Churches, "there is no absolute distinction between eliminating 'defects' and 'improving' heredity. Correction of mental deficiency can move imperceptibly into enhancement of intelligence, and remedies of severe physical disabilities into enhancement of prowess."[16]

Despite the blurry line, 'therapy' seems to garner approval while 'enhancement' is greeted with moral suspicion. Francis Fukuyama speaks for many: "One obvious way to draw red lines is to distinguish between therapy and enhancement, directing research toward the former while putting restrictions on the latter."[17] Geneticist W. French Anderson holds that "we should not step over the line that delineates treatment from enhancement."[18] Both therapy and enhancement require altering our bodies; yet the motive of healing seems distinguishable from that of enhancement.

The line between enhancement and transhumanism can also be blurry. Neruoscientists are currently employing chip lithography to create silicon neuronal chips. These silicon chips have furrows cut in them so that, when implanted in the human brain, actual neurons will grow in such a way that their dendrites and axons interact to create new input and output patterns. The merging of machine and brain could enhance thought, perhaps carrying a person beyond mood alteration or executive improvement to an advance in calculation ability.

Curiously, experiments with computers are going the opposite direction. Because DNA is the single best information storage system yet discovered, scientists are experimenting to see if they can create DNA computers with storage capacities that will leap ahead of anything we have created synthetically. We are technologizing the organic world while organicizing the technological world. Whether inside or outside the human brain, calculation capacity is on the verge of a quantum leap forward.

Bioethicist Paul Wolpe finds these developments astonishing yet promising. "The point of all this is that we really are becoming some kind of cyborg, some kind of posthuman in the sense that for the first

[16] World Council of Churches, Church and Society, *Manipulating Life: Ethical Issues in Genetic Engineering* (Geneva: World Council of Churches, 1982) 7.

[17] F. Fukuyama, *Our Posthuman Future* (New York: Farrar, Strauss, and Giroux, 2002) 208.

[18] W. F. Anderson, "Genetics and Human Malleability, *Hastings Center Report,* 20 (February 1990) 24.

time in history we really are going to incorporate our synthetic technologies into the very physiology of our being – with major, though not necessarily entirely undesirable, consequences."[19] Arthur Caplan denies that these developments will make us post-human; rather, he celebrates their enhancement of our humanity. We should applaud "the new knowledge the brain sciences are providing to try to improve, enhance, and otherwise move toward optimization of our brains."[20]

Ethicists who anticipate developments in neurocognitve enhancement identify four ethical issues.[21] First, safety. What might be the long term biological effects of interventions now via either neurotechnology or psychopharmacology? Will today's Ritalin children tomorrow confront premature cognitive decline? Second, coercion. If neurocognitive enhancement becomes widespread, might we confront situations where people are pressured to enhance their cognitive abilities? Will employers pressure employees, or courts pressure convicts? Third, distributive justice. More than likely, the upper economic classes will benefit more from brain enhancement than those unable to afford such medical opportunities. Is this just? Fourth, personhood and intangible values. Will neurocognitive alterations intersect with what we believe makes us persons? Will it make difficult our ability to envision what it means to be healthy and whole? Will it dull us from appreciating and valuing human life even with its imperfections?

Theology and the Ethics of Enhancement

The challenges of enhancement and transhumanism are knocking on religion's door. Will that door open or remain shut? Heidi Campbell and Mark Walker present transhumanism in terms of its salvific potential. Technology is becoming a rival to religious promise. The transhumanist vision is compelling, they say, because it touches "on a desire for a life that overcomes the brokenness of this world, a place where pain and suffering are eliminated. This is a longing that is articulated in many religious traditions, those that subscribe to a distinctive eschatological belief in a

---

[19] Paul Root Wolpe, "*Neurhotechnology*, Cyborgs, and the Sense of Self," in *Neuroethics: Mapping the Field,* Dana Foundation Conference Proeceedings, San Francisco, May 13–14, 2002 (New York: The Dana Press, 2002) 164.

[20] Arthur Caplan, "No Brainer: Can We Cope with the Ethical Ramifications of New Knowledge of the Human Brain?" Ibid., 97.

[21] Martha J. Farah/Judy Illes/Robert Cook-Deegan/Howard Gardner/Eric Kandel/Patricia King/Eric Parens/Barbara Sahakian/Paul Root Wolpe, "Neurocognitivie Enhancement: What can we do and what should we do?" *Nature Reviews*, 5 (May 2004) 421–425; www.nature.com/reviews/neuro.

future where humanist if perfected and transformed."[22] We are on the brink of a technological eschatology, they assume. What might theologians have to say about this attempt to hand craft our own Omega?

Relatively little has been said by Christian theologians and ethicists on the specific topics of brain enhancement through genetic intervention, neurosurgery, or pharmacology, let alone transhumanism. However, the more general question of enhancement has been raised on occasion. We will look here at the question of therapy versus enhancement in Eastern Orthodox ethics, Roman Catholic moral theology, and Protestant thought.

Eastern Orthodox theologians place their ethics within the framework of eschatological perfection, perfection understood as union with God. "The Christian must place health care, the amelioration of suffering, and the postponement of death within the pursuit of holiness," writes Tristram Engelhardt.[23] Within such a framework, John Breck can celebrate the potential benefit from genetic engineering (GE) for purposes of therapy; but he is more than merely cautious at the prospect of using genetic intervention for purposes of enhancement and eventually eugenics. "The line between therapeutic and eugenic techniques is difficult to draw," he writes. "While theoretically it is possible for GE to replace *homo sapiens* with a superior *homo novus*, such 'enhancement technology', is far in the future ... Nevertheless, the potential for 'improving' on God's blueprint for human life is such that serious ethical questions must be addressed here and now, by the Churches as well as by public and private regulatory bodies."[24] If the Orthodox are cautious about biological enhancement, they are likely to be even more reticent when confronted with transhumanism.

Roman Catholic theologians are also less than likely to advance an ethic supporting enhancement or posthumanism. A recent International Theological Commission working for the Vatican's Congregation for the Doctrine of the Faith has published a thorough discussion of Christian anthropology, *Communion and Stewardship: Human Persons Created in the Image of God*. It celebrates what it calls the "biogenetic characteristics" that allegedly make each person unique. To modify a person's genetic code, regardless of motive, would infringe on a person's identity, says the commission. "Changing the genetic identity of man as a human person through the production of an infrahuman being is radi-

[22] Campbell/Walker, "Religion and Transhumanism," 2.

[23] H. Tristram Englehardt, Jr., *The Foundations of Christian Bioethics* (Liss: Swets & Zeitlinger, 2000) 354.

[24] J. Breck, *The Sacred Gift of Life: Orthodox Christianity and Bioethics* Crestwood NY: St. Vladimir's Seminary Press, 1998) 195.

cally immoral. The use of genetic modification to yield a superhuman or being with essentially new spiritual faculties is unthinkable, given that the spiritual life principle of man – forming the matter into the body of the human person – is not a product of human hands and is not subject to genetic engineering."[25] The logic seems to be this: genetically designing a posthuman is immoral because it is impossible and, if it were possible, it would still be immoral because it compromises human identity.

Significant here is that Catholic caution regarding enhancement would arise not from the technologies themselves, but from an understanding of what human excellence means. Catholics are leery of using medicine for enhancement because efforts at enhancement are so obviously bound up with value choices. A Protestant writing in a Catholic journal, Donal O'Mathuna, says "In making a person taller medicine promotes the belief that short people are of lesser value and that height is significant in achieving the good life."[26] Some of the values which might guide enhancement are questionable from a Christian perspective: "Physical health is not the ultimate priority in Christian eyes."[27] In short, if the *telos* or proper end of human life is the development of a relationship with God and the cultivation of those excellences or virtues necessary for living in community, then even some forms of suffering can be seen as serving those ends. Not all suffering should automatically be removed from human life. "Attempting to remove the trials and difficulties of life by genetic enhancement might derail the very ways in which God wants to shape our characters."[28]

Is there wiggle room here? Pope John Paul II did not explicitly forbid genetic enhancement. What needs to be protected is human identity and dignity; and if enhancement could render this protection perhaps it would not be *prima facie* illicit. James Walter comments: "experiments that are not strictly directed toward therapy but are aimed at improving the human biological condition (enhancement) can be justified, at least in part, on the grounds that the experiments respect the human person by safeguarding the identity of the person as one in body and soul (*corpore et anima unus*)."[29]

---

[25] *Communion and Stewardship: Human Persons Created in the Image of God*, 91. The Vatican, International Theological Commission, Congregation for the Doctrine of the Faith. http://www.vatican.va/roman_curia/congregations/cfaith/cti_documents/rc_con_cfaith_doc.

[26] Donal P. O'Mathuna, "Genetic Technology, Enhancement, and Christian Values," *The National Catholic Bioethics Quarterly*, 2:2 (Summer 2002) 283.

[27] Mathuna, "Genetic Technology," 284.

[28] Ibid., 295.

[29] James J. Walter, "Catholic Reflections on the Human Genome," *The National Catholic bioethics Quarterly,* 3:2: 275–286 (Summer 2003) 281.

Just how much should we expect from enhancement or related improvements? Might we become unrealistic? Celia Deane-Drummond is skeptical about the use of "salvific language" on the part of scientists pressing for genetic enhancement of human intelligence.[30] Too much is being promised here. Not only are expectations raised above what reality can possibly deliver; extravagant promises rely on values that are impatient with our humanity, that take us beyond our natural human state. What Deane-Drummond would propose is a virtue ethic drawn from an appropriation of wisdom. "The theological motif we need to recover as an appropriate response to the new biology is that of wisdom. Such wisdom ... offers guidance about life based on the practical historical reality stemming from everyday problems and issues."[31] Wisdom "would encourage caution on the part of humans rather than impulsive action."[32] Caution and prudence are called for, not prohibition.

Turning to the Protestants, American evangelical bioethicist, C. Ben Mitchel, sets the issue up this way: "I want to reserve healing for restoration of lost capacities, and enhancement for improving on the species, or on the statistical norm."[33] Interestingly, we find within American evangelical circles attempts to connect the blurry line between therapy and enhancement with the fall in the Garden of Eden. Would medical interventions that seek to restore what was lost in the fall count as therapy? Would the lengthening of life or even the overcoming of death count as therapy; or would it count as enhancement? This is not easy to sort out theologically. "If, for instance, one holds that physical death and aging are a result of the Fall, then it would appear that a genetic battle against aging and death could be justified, and therefore qualify as therapy ... However, more recent shifts in theological thinking, not immune to the profound influence of evolutionary theory where both aging and death are an *integral* part of how we humans have come to be, have resulted in a rejection of the idea that physical death was introduced into the world by Adam's sin."[34] In the latter case, life extension through genetic engineering would count as enhancement.

Several centrist Protestant ethicists note how difficult it is to draw the line between therapy and enhancement. Yet, some ethicists still believe it is possible to specify a base-line of human functioning that

---

[30] Deane-Drummond, *Brave New World?*, 29.

[31] Celia E. Deane-Drummond, *Creation Through Wisdom: Theology and the New Biology* (Edinburgh: T&T Clark, 2000) 234.

[32] Celia E. Deane-Drummond, *Biology and Theology Today* (London: SCM Press, 2001) 117.

[33] C. Ben Mitchel, "Define 'Better'," *Christianity Today,* 48:1 (January 2004) 43.

[34] Todd T. Daly, "Therapy vs. Enhancement: The Problem posed by Anti-Aging Technologies," 2005, The Center for Bioethics and Culture Network, http://www.thecbc.org/redesigned/research_display.php?id=199.

is part of the intended order of creation. When someone falls below this base-line, genetic interventions are called therapy or correction. When an intervention would move someone above the base-line, it is called enhancement. Paul Ramsey has argued strenuously for an "exact" and limited meaning of the term "genetic therapy," precisely so that it could not be used for enhancements that treat only desires.[35] If we would intervene in our own genetic code to improve on human nature, we would usurp what rightfully belongs solely to God. We would be playing God. Ramsey is remembered for having penned these lines in 1970: "Men ought not to play God before they learn to be men, and after they have learned to be men they will not play God."[36] Such an action would constitute human pride or *hubris*.[37]

LeRoy Walters and Julie Gage Palmer have asserted that "disease and disability" are "evils that should be overcome as quickly and efficiently as possible."[38] While all short children may experience pain or discrimination, the question of whether there is a physiological basis for the short stature is taken to be morally relevant for policy purposes. "We are attempting to draw a sharp line between *bona fide* illness ... and physical traits that can lead to discouragement or discrimination or both ..."[39] They approve genetic enhancement for children of short stature who have hormone deficiencies, but not for children of short stature who do not have hormone deficiencies.

James C. Peterson is one of the most thoughtful Protestants to take up the issue of enhancement. The line between therapy and enhancement is blurry, he admits. "The problem is both with the indistinctness of the line and with the rationale for holding it. There is no adequate conceptual distinction between cure of disease and enhancement of capacity that would allow us to make a principled argument for cure of disease that would not over time also allow genetic intervention for enhancement."[40] Peterson works with an anthropology and ethic of creative love. "If human beings are called to develop themselves, pur-

---

[35] Paul Ramsey, "Genetic Therapy: A Theologian's Response," in Michael Hamilton, ed., *The New Genetics and the Future of Man* (Grand Rapids: Wm. B. Eerdmans 1972).

[36] Paul Ramsey, *Fabricated Man: The Ethics of Genetic Control* (New Haven: Yale University Press, 1970) 55.

[37] For an analysis of the concept of "playing God," see Ted Peters, *Playing God? Genetic Determinism and Human Freedom* (London/New York: Routledge, [2]2002).

[38] Leroy Walters/Julie Gage Palmer, *The Ethics of Human Gene Therapy* (Oxford/New York: Oxford University, 1997).

[39] Ibid., 113.

[40] James C. Peterson, *Genetic Turning Points: The Ethics of Human Genetic Intervention* (Grand Rapids MI: Wm. B. Eerdmans, 2001) 252.

poseful and direct enhancement of capacity could be appropriate, or according to some even required."[41]

Transhumanism and the Question of Embodied Identity

All the above theological discussion pertains to the human body. We find theologians making two assumptions regarding our identity. First, who we are as a person is embodied. Second, changes in the body, even if resulting in changes in the mind, do not risk a loss of identity. Beyond therapy and even beyond enhancement, our transformed self will still be our self. When we travel the path from therapy and enhancement toward transhumanism, however, these two assumptions do not travel with us. Theologians will eventually need to consider the implications of neuroscience aiming at transhumanism.

Curiously, the assumption at work in transhumanism is that human intelligence and hence human personhood can become disembodied. This is by no means a return to substance dualism and belief in a metaphysical or spiritual entity; yet, it is still curiously akin to pre-modern beliefs in disembodied intelligent existence. Information patterns have simply replaced the disembodied soul.

Old fashioned body lovers object. "Modern transhumanism is a statement of disappointment," writes Brian Alexander. "Transhumans regard our bodies as sadly inadequate, limited by our physiognomy, which restricts our brain power, our strength and, worst of all, our life span. Transcendence will not be found in the murky afterlife of the usual religions, but in technological and biological improvement."[42]

Objections come from at least two allied directions, from science and from theology. Scientifically, we are becoming increasingly convinced that brains and hence minds are embodied,[43] perhaps even communal. Despite the significant role played by our genetic inheritance, we are learning that during our childhood years our brain functioning is itself developing; environmental factors have a decisive impact on brain development as well as formation of the human self. Brains, bodies, and

---

[41] Ibid., 288.

[42] Brian Alexander, *Rapture: How Biotech Became the New Religion* (New York: Basic Books, 2003, 51.

[43] "I believe that the final Copernican revolution will be completed within the next century," writes Terrence Deacon; "with the explanation of how the electrical-chemical processes in our brains produce the experience of being persons." "How I Gave Up the Ghost and Learned to Love Evolution," in *When World's Converge*, ed. by Clifford N. Matthew/Mary Evelyn Tucker,/Philip Hefner (Chicago/LaSalle IL: Open Court, 2002) 137.

the environment are in constant interaction. The transhuman "proposals are based on the confidence that the human mind can be explained wholly by a computational functionalist approach," is the assessment of Gregory Peterson. "While there are important reasons to consider such a possibility, it is by no means certain … Is a downloaded version of me the same me? … We have become increasingly aware that the mind is intimately connected to physical states. We are not simply disembodied reasoning machines but persons in a bodily and communal context."[44]

Christian theology during the modern period has reaffirmed the goodness of the body and jettisoned the substance dualism of the past. "Transhumanism is in some ways a new incarnation of Gnosticism," complains Mayo Clinic hematologist and bioethicist, C. Christopher Hook; "It sees the body as simply the first prosthesis we all learn to manipulate."[45] Hook goes on to declare that "As Christians, we have long rejected the Gnostic claims that the human body is evil. Embodiment is fundamental to our identity, designed by God, and sanctified by the Incarnation and bodily resurrection of our Lord."[46] This theological affirmation of embodied human existence suggests unqualified commitment to therapy, perhaps commitment to enhancement, and skepticism about transhumanism.

Cybernetic Immortality?

What if a transhumanist would provide us with a new body, a simulated replica of our present body for the use of our pattern of information processing? Could we live forever? Would this meet the theological concern for embodiment?

Key to opening the transhuman door to the future is the development of Artificial Intelligence (AI). The link between AI and the transhuman vision is a series of assumptions: (1) what makes us human is our intelligence; (2) intelligence consists of information processing; and (3) the transfer of the pattern for information processing from our brain to a machine is feasible. Further, if we keep the machines in perpetual repair, we can live for ever. We will have achieved cybernetic immortality.

Such a promise tantalizes. Tulane physicist Frank Tipler combines AI and cosmology to project an image of a future state of intelligent

---

[44] Gregory R. Peterson, *Minding God: Theology and the Cognitive Sciences* (Minneapolis: Fortress Press, 2003) 218.

[45] C. Christopher Hook, "The Techno Sapiens are Coming," *Christianity Today*, 48:1 (January 2004) 38.

[46] Ibid.

existence that constitutes immortality. We will create computers in our image, and these computers in turn will create a replica of ourselves; and this replica will persist beyond our death and beyond the death of the universe.

Tipler's method unfolds the religious candy wrapper while setting aside the religious candy inside. Even though he mentions the Easter resurrection of Jesus and the eschatological writings of Paul, he tries to promulgate a strictly secular immortality for techno-sapiens.[47] Tipler bases his promise on the history of evolutionary progress. Resurrection here will be the result of a future evolutionary event in which life understood as information processing will take hold of its own destiny and create a supra-biological environment for its existence just prior to the moment when the physical world self-destructs.[48]

Tipler addreses a couple of concerns theologians would likely raise here: perfection and identity. First, he is sensitive to the human yearning for perfection. Our present state of existence is not satisfactory. We do not hunger simply for life beyond death. We hunger for salvation. So, without using the term 'salvation', Tipler announces that the perpetuation of our pattern of information processing within our simulated body will transcend the previous model by eliminating bodily defects such as missing limbs; youth will be substituted for old age; sight for blindness; etc.

The second concern is whether continuity of identity will be maintained. Tipler answers, yes. Anticipating objections that total death followed by total re-creation denies continuity, Tipler responsively argues that continuity in conscious self-identity is both necessary and possible. Replication is not annihilation. To be resurrected as a replica of one's former self does not deny that it is the same self. The identity of the information patterns within which we are aware of our experience of the world and ourselves seems to be sufficient for Tipler. In sum, what is resurrected is the immaterial form but not the material substance of who we presently are.

> An exact replica of ourselves is being simulated in the computer minds of the far future. This simulation of people who are long dead is "resurrection" only if we adopt what philosophers call the "pattern identity theory"; that is, the essence of identity of two entities which exist at different times lies in the (sufficiently close) identity of their patterns. Physical continuity is irrelevant.[49]

---

[47] Frank J. Tipler, *The Physics of Immortality* (New York: Doubleday, 1994) 305; 309–313.

[48] Tipler, *Physics of Immortality*, 225.

[49] Ibid., 227.

Is this pattern theory merely another example of platonic body-soul dualism in which a nonmaterial soul is extracted permanently from a material base? No, says Tipler. Tipler's simulation so emulates the physical body that, for all practical purposes, what resurrected souls experience is physically real. According to Tipler, resurrected souls experience themselves in their environment; and this environment is experienced as if it were physical. In a surprising move, Tipler reiterates Bishop William Berkeley's subjective idealism: to be is to be perceived.[50] If as a computer simulation we perceive physicality, the physicality exists thereby.

Let me ask a theological question: is the replication of a pattern of information processing sufficient? Let me be sympathetic for just a moment. Rather than requiring God to locate and piece together all the molecules of our previous body, could God's task at the resurrection be simplified by merely reassembling the pattern of our body's molecules? Rather than the molecules as matter, might we say that the form is what counts? Could God get by with merely remembering and reincarnating our form or pattern? Tipler is reminiscent of Origen: "The previous form does not disappear, even if its transition to the more glorious occurs ... although the form is saved, we are going to put away nearly [every] earthly quality in the resurrection ... [for] 'flesh and blood cannot inherit the kingdom' (1 Cor. 15: 50)."[51] For Origin and Tipler the form is saved, but not the substance. Is this sufficient to guarantee continuity of personal identity? I doubt it. Our identity is the product of our biography. Our identity is something gained over time, inclusive of scars embedded in our limbs combined with our remembered thoughts. Is there not more to our individual consciousness than information processing, namely, our own information processing informed and influenced by the accumulated history of our bodily functions?

Let me ask an additional question: if the key to cybernetic immortality is replication of the soul's pattern, what would happen in the event of multiple replications? We already know what computer clones are. In principle, the soul's pattern could be replicated many times, not just once. Which pattern would maintain the individual's identity? Does continuity of unique personal identity require some degree of substance uniqueness?

---

[50] Tipler, *Physics of Immortality*, 211.

[51] Origen, "Fragment on Psalm 1: 5," cited by Carolyn Walker Bynum, *The Resurrection of the Body in Western Christianity, 200–1336* (New York: Columbia University Press, 1995) 64. My original analysis of Tipler can be found in Ted Peters, "Resurrection: The Conceptual Challenge," in *Resurrection: Theological and Scientific Assessments*, ed. by Ted Peters/Robert John Russell/Michael Welker (Grand Rapids MI: Wm. B. Eerdmans, 2002) 297–321.

Just how realistic is the Tipler theory, scientifically speaking? Not very, either physically or biologically. John Polkinghorne dubs these ideas as "excessively speculative in the assumptions that they make about physical processes in unexplored circumstances, particularly in Tipler's case. The closing instants of a collapsing universe involve physical processes at energies vastly in excess of those of any regime of which we could claim to have an understanding ... The speculations of the physical eschatologists are also chillingly reductionist in tone. Life is equated to the mere processing of information. Only if one believes that humans are no more than computers made of meat could one regard their replacement by computers made of bizarre states of matter as affording a picture of continuing fulfillment."[52]

Just how realistic is cybernetic immortality, philosophically speaking? Although the late philosopher Paul Ricoeur does not address Tipler directly, he says something quite relevant. The very concept of the brain and the very work of neuroscience are products of the human mind. This means the mind has sufficient independence so as to authorize its own self-study, something the brain could not do on its own, to be sure. "I propose we adopt the term *substrate* to denote the relation of the body-as-object to the body as it is experienced, and therefore of the brain to the mental."[53] By *substrate* Ricoeur means something like Aristotle's material cause. Ricoeur makes this move to challenge the unfounded assumptions of the scientists that the brain is the efficient cause of human mental operations. Neuroscientists over extend themselves, complains Ricoeur. At best, we can identify a correlation between brain activity with mental activity; but no warrant exists for a causative relationship. "The brain is the substrate of thought (in the broadest sense of the term) and ... thought is the indication of an underlying neuronal structure. Substrate and indication would thus constitute the two aspects of a dual relation, or correlation."[54] Ricoeur puts a fence up against reducing human consciousness and mental activities to the pattern of neuronal functioning. The implication is that even if the pattern of brain function could be replicated, there is no guarantee – not even a likelihood – that the consciousness of the person would be replicated with it.

Just how realistic is cybernetic immortality, theologically speaking? Noreen Herzfeld is critical. "The assumptions regarding the natures of the human person and of eternal life that underlie the hope of an

---

[52] John Polkinghorne/Michael Welker, ed., *The End of the World and the Ends of God* (Harrisburg PA: Trinity Press International, 2000) 33.

[53] Jean-Pierre Changeux/Paul Ricoeur, *What Makes Us Think?*, tr. by M.B. DeBevoise (Princeton: Princteon University Press, 2000) 46.

[54] Changeux/Ricoeur, *What Makes Us Think?*, 47.

immortal presence within computers are quite different from those of most Christians. Cybernetic immortality assumes a dualistic understanding of the human person, a conception of eternity as 'a long time', and a hubristic faith in human power."[55] Cybernetic immortality cannot become a substitute for what Christians understand as resurrection of the body.

What are the ethical implications? Denigration of the body. Herzfeld continues: "Our finite bodies are an integral part of who we are. The essential nature of the human being always contains two inseparable elements, self-transcending mind and finite creaturely being. The denial of the latter has led to a denigration of both women and the natural environment. Cybernetic immortality leads directly into these twin denigrations … It is notable that cybernetic immortality had been suggested as a possibility only in the writings of rich, white males."[56]

The cybernetic immortality projected by Frank Tipler represents transhumanism in its most audacious and extreme form. Much more modest forms of integrating human brains and computational machines could fit within the broader categories of therapy or, morel likely, enhancement. Instead of thinking of cyborg brains as transhuman, we could think of them simply as fabulously human.

Eschatological Transformation and Perfected Human Beings

"Render therefore unto Caesar the things which are Caesar's; and unto God the things that are God's," says Jesus (Matthew 22:21).[57] In our case, Caesar is science. Any ethic should begin by distinguishing realistically what belongs in the domain of science from what belongs in the domain of divine promise. Once this distinction has been made, then we can look for the connection. God has promised an eschatological transformation, the Omega, the new creation. Our task this side of Omega is to engage in the much more modest work of transformation in order to improve human health and flourishing.

The eschatological orientation seeks the good in the future, not the past. It presumes the world anticipates its own betterment, that our human condition yearns for redemption. Human enterprises such as scientific research and medical advance contribute to healing and overall human well being. They are attuned to the work of our God of

[55] Herzfeld, *In Our Image*, 73.
[56] Ibid., 74.
[57] New King James Version 1982.

transformation who promises to make all things new. Under the right conditions, science can be considered a godly vocation.

I like to speak of an eschatologically oriented ethic as a "proleptic ethic," with 'prolepsis' indicating that today's actions anticipate tomorrow's transformation.[58] A proleptic ethic begins with a vision, a vision of the perfected human being residing in a new creation. "And God shall wipe away all tears from their eyes; and there shall be no more death, neither sorrow, nor crying, neither shall there be any more pain: for the former things are passed away" (Revelation 21:4, KJV). The Bible's closing apocalypse envisions a total healing. Scientific and medical contributions to human betterment today, modest as they may be, anticipate, and thereby participate in, God's final redemption.

The task of the ethicist, I believe, is to devise middle axioms that connect the grand eschatological vision of a new creation with our quite human responsibilities in the present time. Middle axioms would be theological principles providing ethical support for scientific research and medical technology that contribute to improved therapy and perhaps some degree of enhancement. Theologically, we need to affirm that scientific and technological transformation actually participate in the renewing of our world in a way that is both human and divine. Yet, blanket baptism of anything and everything new would be imprudent, to say the least. Constructing such middle axioms requires wisdom and prudence, what the Aristotelian tradition knows as *phronesis* and hermeneutical philosophers know as *applicatio.* Such application incorporates the unavoidable ambiguity inherent in assessing the practical outcomes of speculative proposals. The task of the ethicist remains: face the ambiguities, invoke wisdom, think prudently, and render the best judgment that finite considerations can produce.

Enhancement and Justice

As we turn to our three areas of medical research – therapy, enhancement, and transhumanism – my own considered judgment is the following. Therapy or healing is incontrovertibly a divinely appointed ministry as well as a humane enterprise. Transhumanism can be dismissed because it is scientifically and philosophically unrealistic as well as theologically and ethically misdirected. When it comes to enhancement, however, we confront complexities and nuances that may require additional analysis.

---

[58] Ted Peters, *GOD – The World's Future: Systematic Theology for a New Era* (Minneapolis: Fortress Press, ^{2}2000) chapter 12.

When we ponder the standard definition of 'enhancement', we must take note of the competition component. By definition, enhancement seems to require that one individual become superior to others, that one person stands out as being above average. Is enhancement necessarily anti-egalitarian? Could widespread enhancement technologies lead to a new eugenics movement, in which some segments of society benefit and others suffer discrimination? If so, then we must face the justice issue.

In the United States, the National Council of Churches has raised cautions about eugenic programs.[59] Similarly, the United Church of Christ in 1989 welcomed the development of genetic engineering *provided* there would be appropriate regulation and "justice in distribution."[60] The United Church of Canada expressed as a general principle that the "rights of the weaker and the needy" must be protected in any genetic interventions.[61] Karen Lebacqz has also warned that expensive advances in medical genetics may be "no deal for the poor."[62]

These observations regarding justice uncover the heart of the problem with enhancement in all its forms. If enhancement comes to mean that the capacity of one individual is raised to a level superior to others, the question of justice arises. For the near and medium range future, medical technologies will be expensive. Only some segments of society will be able to afford such medical advances. The financially rich may also become the "genrich," to use the language of Lee Silver.[63]

"Justice is concerned broadly with the idea that each is given her or his due," writes Deane-Drummond. "Unlike many other virtues, justice specifically governs relationships with others."[64] The aspect of enhancement which draws our ethical scrutiny here is its relationship with others. The risk or threat is not found in improving one's memory or executive capabilities; the justice problem arises due to the relative social status of such a person once this has been accomplished. The threat of injustice is found in the likelihood that privileges will go to the enhanced; and the unenhanced will suffer marginalization.

---

[59] Ronald Cole-Turner, *New Genesis: Theology and the Genetic Revolution* (Louisville: Westminster John Knox Press, 1993) 73.

[60] Cole-Turner, *New Genetics*, 76.

[61] Audrey Chapman, *Unprecedented Choices* (Minneapolis: Fortress Press, 1999) 60.

[62] Karen Lebacqz, "Fair Shares: Is the Genome Project Just?" and "Genetic Privacy: No Deal for the Poor," in Ted Peters, ed., *Genetics: Issues of Social Justice* (Cleveland: The Pilgrim Press 1997) 239–254.

[63] See: L.M. Silver, *Remaking Eden: How Genetic Engineering and Cloning Will Transform the American Family* (New York: Avon, 1998).

[64] Deane-Drummond, *Ethics of Nature,* 15.

Professional sports are currently rocked with scandal due to drug induced physical enhancements.[65] Intuitively, the public knows this is unfair, unjust. We can expect to see legal rules put in place to keep the competition egalitarian. Might such policies be devised to govern genetic and neurocognitive enhancements? Enhancement technology could be parsed accordingly. Does enhancement necessarily lead to social inequality?

Now, we might reverse the logic for a moment. Suppose enhancement for some and not others does in fact lead to social inequality. Let us pose some other justice questions. Should we deny to an individual the opportunity for neurocognitive enhancement simply because others in the community might not have equal opportunity? Should the fear of social or economic injustice be sufficient to warrant denial of increased well being for an individual? Such questions require attention by this generation of theologically guided ethicists.

In light of this, would it be prudent to alter our understanding of 'enhancement' and, thereby, our understanding of the perfect human being? Could we eliminate the social competition component? Rather than work with a standard base line or human average with regard to memory or executive abilities, could we simply ask what enhancement could mean on a case by case basis? Could we think in terms of incremental improvements in individual well being? That is to say, could we devise an ethical principle that would exclude enhancement for competitive purposes but encourage it for individual flourishing? Again, these are questions deserving of attention.

Conclusion

In summary, here we have followed the track of therapy, enhancement, and transhumanism en route to neurocognitive improvement through genetic engineering, neuro-technology, and pharmacology. We have assessed the direction this is going in light of a proleptic ethic buttressed by a virtue ethic that emphasizes wisdom and prudence. We have found that the concept of therapy would welcome any advances in neuroscience that could lead to healing of brain and mind. We have found that transhumanism is founded on unsubstantiated assumptions regarding the human person and human intelligence; and that theolog-

[65] American baseball player Jose Canseco describes himself as "a nearly superhuman athlete" due to his use of steroids and growth hormone. *Juiced* (San Francisco: HarperCollins, ReganBooks, 2005) 2.

ically transhumanism confuses ultimate transformation with the more modest possibilities of genuine improvement in human well being.

Finally, we have followed the track of the enhancement train. With regard to the individual, enhancement of mood, memory, or executive capabilities in themselves could lead to an improvement in human well being without raising ethical objections. However, ethical problems arise when enhancement is used to create a fabulous human being who is superior to others in the community and who benefits unjustly from this superiority.

It is my own considered judgment that the risk of injustice is insufficient to slow down or curtail laboratory research on improving neurocognitive capacities. The task of the ethicist should be applied to distribution of the medical benefits so as to meet the standards of social and economic justice.

Chapter 7

The Post-Human Soul, Part Two

"Do not fear those who kill the body but cannot kill the soul," warns Jesus; "rather, fear him who can destroy both soul and body in hell" (Matthew 10:28). Jesus is worried that we might lose our soul. But, just finding our soul in the first place is quite a chore in our materialistic and religiously relativistic age. The materialism of our era buries the soul so deeply in our genes or in our brain's biology that we can hardly dig it up any more; and the deluge of religious options in our pluralistic culture makes us wonder which if any we should take seriously. Just what is the soul that we might lose, or find?[1]

The question of the human soul is subject to pressures both external and internal to Christian anthropology. The external pressures come from new studies in the neurosciences that connect brain function with mind function, and from trans-humanism which envisions a post-human future complete with cybernetic immortality. "There is interdependence between what is happening at the physical level of brain processes and at the levels of cognition and behavior," writes British neuroscientist Malcolm Jeeves. Neruoscientific research signals that the rate of progress in "ever-tightening links between mind and brain continues to accelerate."[2]

In the third century Tertullian pitted the philosophy of Athens and the theology of Jerusalem against one another in an uneasy dialogue. Today, says Gregory Peterson, we need to place theology into dialogue with neuroscience and cognitive science. "If Tertullian were alive today, he might not contrast Jerusalem with Athens as much as, say, with Los Alamos or Fermilabs."[3] When directing our attention in this direction, here is what we see: "all mental processes will be understood in terms

---

[1] This chapter is a revision of Ted Peters "The Soul of Trans-Humanism," *Dialog* 44:4 (Winter 2005) 381–395.

[2] Malcolm Jeeves, "Human Nature: An Integrated Picture," in *What About the Soul? Neuroscience and Christian Anthropology*, ed. by Joel B. Green (Nashville: Abindgon, 2004) 173.

[3] Gregory R. Peterson, *Minding God: Theology and the Cognitive Sciences* (Minneapolis: Fortress Press, 2003) 6.

of the functioning of the brain. Thus, according to most experts, our mental abilities do not require the existence of a supernatural soul."[4] This is the external stimulus. What will be the theological response?

Internally, theologians for some decades now have been retrieving the positive biblical value of the human body along with the physical world; and they have internalized the disquisitions of philosophers of religion on the mind-body problem. The result is a contemporary Christian anthropology that celebrates our physical nature and emphasizes a relational and integrated understanding of human selfhood and personhood. "Personhood," writes Wolfhart Pannenberg, "transcends all the singularities and changes of circumstances because it finally draws upon the relation to God as the source of its integrity."[5] Now, we ask: what direction might this internal reflection on Christian anthropology take us regarding the nature of the human soul?

In what follows in this chapter, we will ask just what Christian theology needs to say about the human soul in response to the stimulations of genetic science, neuroscience or cognitive science, and transhumanism. The provocation comes from two directions simultaneously, from the science that analyzes human nature and from the technology that could alter human nature.

It is my own contention that human nature is not yet fully established. It remains a promise, a divine promise. The Easter resurrection of Jesus prefigures what we human beings will become, eschatologically. When science and technology of the type we will look at here seems to evoke reflection on what Christian theology means by the human soul, we need to orient this reflection toward our ultimate future, God's promised omega.

Neuroscience and Cognitive Science as Theological Stimulus

The science of the brain and the technology of the brain, curiously, are dysfunctional. They are contradictory, at least in part. The tendency in the science is toward reductionism, toward reducing our minds and our souls to biological activity. The contrary tendency in transhumanist technology is to view the mind or soul as immaterial, as something that can become dis-embodied and re-embodied. The first

[4] Jerome W. Elbert, "Does the Soul Exist? From the Mythological Soul to the Conscious Brain," in *Science and Religion: Are They Compatible?*, ed. by Paul Kurtz (Buffalo NY: Prometheus Books, 2003) 249.

[5] Wolfhart Pannenberg, *Systematic Theology,* tr. by Geoffrey W. Bromily (3 Volumes: Grand Rapids MI: Wm. B. Eerdmans, 1991–1998) II:200.

tends toward a substance monism or materialism, the second toward a substance dualism. One implication is this: theologians cannot allow science to do their work for them. They will need to turn to their own internal resources to develop a Christian anthropology that is informed by the science but not dictated by the science.

We will begin with the reductionist stimulus. "Neuroscientists are rapidly learning to read and mold the human brain, and to predict behavior and disease well into the future. Meanwhile, bioethicists at Stanford and elsewhere are trying to keep pace by anticipating the potential landmines. Those mines threaten to detonate across a broad array of sectors – from schoolroom to courtroom, hospital to voting booth, homeland security to human rights."[6] Perhaps what this Stanford bragging deems to be a landmine is materialism, the reduction of what people think is their mind to its physical substrate. If you think you have a spiritual dimension and are told you are nothing but physical, it might ruin your day.

Neuroscience is giving birth to *neurotheology.*[7] By *neuroscience* we refer to the scientific study of the nervous system and its functions. This includes brain research. Bridge builders between neuroscience and religion believe they can provide a chemical explanation of religious experience. The research includes what happens when people engage in meditation and worship. They argue that the sense of diminishment of self and an awareness of oneness with the divine that is experienced by people during transcendental meditation is associated with diminished activity in the parietal lobes of the brain. Religious experience understood as a sense of absolute unitary consciousness can be explained as brain activity.[8] "Such experiences may be described as a perfect union with God, as in the *unio mystica* of the Christian tradition, or else the perfect manifestation of God in the Hindu tradition," writes Andrew Newberg.[9] The goal here is to bring religious discourse under one rational explanatory scheme, namely, the scheme of neuroscience.

---

[6] Joan O'C. Hamilton, "If They Could Read Your Mind," *Stanford* (January/February 2004) 38.

[7] The term *neurotheology* seems to have been coined by James B. Ashbrook, "Neurotheology: The Working Brain and the Work of Theology," *Zygon* 19 (1984) 331–350. See: Palmyre Oomen, "Neurotheology," *Encyclopedia of Science and Religion*, J. Wentzel van Huyssteen, editor in chief (2 Volumes: New York: Macmillan, Thomson, Gale, 2003). II:617–618.

[8] See: Eugene D'Aquilli and Andrew Newburg, *The Mystical Mind: Probing the Biology of Religious Experience* (Minneapolis: Fortress, 1999).

[9] Andrew Newberg, "Experience, Religious: Cognitive and Neurophysiological Aspects," *Encyclopedia of Science and Religion*, I:309.

The category of *cognitive science* could be a broader one, although it is sometimes used interchangeably with neuroscience. Michael Arbib says cognitive science has a twofold aim: "to enhance our understanding of human thought and behavior and to provide new strategies for building intelligent machines ... [C]all us cyberneticians, model builders, brain theorists, cognitive or computational neuroscientists ..."[10] This field traces the path of perception through interpretive filters that are said to be supplied by brain structure. It is a biological form of philosophical hermeneutics. The difference is that what philosophers previously thought was the product of language and tradition is here connected to chemical processes in the brain.

Despite the friendly smiles of the neuroscientists and the cognitive scientists, these fields threaten classical religious thinking because of biological reductionism. What we previously thought was spiritual or mental or transcendental is here reduced to what is bodily, physical, and chemical. Nobel Prize winner Francis Crick throws down the reductionist gauntlet with his oft quoted remark: "'You', your joys and your sorrows, your memories and your ambitions, your sense of personal identity and free will, are in fact no more than the behavior of a vast assembly of nerve cells and their associated molecules."[11]

Mary Midgley is sharply critical of Crick's reductionism. She accuses Crick of ideological materialism, of arbitrarily dropping the immaterial dimension of consciousness. "Amputation of the mind is to be performed by reductive techniques, translating whatever needs to be said about minds into statements about bodies."[12] What is important to observe here is that this reductionism is actually a philosophical move. In itself, it is not warranted by the science. What we see as cognitive science or neurotheology is actually an ideological overlay. The case for reducing mental activity including religious consciousness to brain function has not actually been made. The empirical evidence is not yet in. What passes for a scientific challenge is still at this stage only an ideological challenge, the ideology of reductionist materialism if not eliminative materialism.

Be that as it may, contemporary Christian anthropology dare not numb itself to the scientific stimulus. "Issues of reduction and auton-

[10] Michael Arbib, *The Metaphorical Brain 2: Neural Networks and Beyond* (New York: John Wiley & Sons, 1989) 4.

[11] Francis Crick, *The Astonishing Hypothesis: The Scientific Search for the Soul* (New York: Scribner's Sons, 1994) 3.

[12] Mary Midgley, "The Soul's Successors: Philosophy and the Body," in *Religion and the Body*, ed. by Sarah Coakley (Cambridge: Cambridge University Press, 1997) 53.

omy strike at the core of what theology is about," is the challenge as Gregory Peterson sees it.[13]

Just what kind of response theology might make to the neuroscience stimulus is multiple. Philip Clayton lists the alternative responses by theology: (1) *the Arbib Credo*, which is the extreme form of reducing all theological insights about the human person into neuroscientific terms; (2) *watch-out-ism*, which says theologians should watch out because eventually neuroscience will explain everything religious in terms of brain activity; (3) *soul-based explanations*, which will attempt to immunize Christian anthropology from science by brute appeal to classical theologies of the soul; (4) *instrumentalism and agnosticism*, which sees brain research as indispensable to the advance of scientific understanding while remaining agnostic on religious implications; (5) the *no conflict* view, which believes changes made in theology will quickly overcome the conflicts with science; and (6) *compatibilism*, which holds that the results of neuroscience neither prove theology nor disprove it; so this science is compatible with theology.[14]

The reductionism of neurotheology combined with the dualism of posthumanism will be the challenge we take up in what follows. We will be critical of the Arbib Credo approach on the grounds that what appears to be a scientific challenge may in fact be a strictly philosophical or even ideological challenge. Perhaps our approach here will combine soul-based explanations with the no-conflict option, but doing so without the sense of siege or timid fear of science that might be implied in Clayton's nomenclature.

Now we are ready to jump fully into the concept of transhumanism, known to some as the *post-human future*. Here, we must confront a most intransigent assumption, namely, who we are as persons is centered in brain activity. In addition, transhumanism assumes that intelligence as a pattern of brain activity is the home of our personhood. It is the cognitive code or information pattern that is definitive. We are what our brains think, so to speak. Therefore, to improve the human situation we must improve on biological function insofar is it enhances cognitive function; and, if necessary, we might even transfer cognitive function into a machine. The goal of transhumanism is to download the contents of human consciousness onto a vast computer network and, via the network, achieve a kind of disembodied yet intelligent im-

---

[13] Peterson, *Minding God*, 19.

[14] Philip Clayton, "Neuroscience, the Person, and God: An Emergentist Account," in *Neuroscience and the Person: Scientific Perspectives on Divine Action,* ed. by Robert John Russell/Nancey Murphy/Theo C. Meyering/Michael A. Arbib (Vatican City State/Berkeley: Vatican Observatory and CTNS, 1999) 185–186.

mortality. As a machine, we could enjoy perfections that the limits of biology prevent us from attaining.

In the previous chapter we outlined what the scenario leading toward a post-human future looks like to a transhumanist. At stage one, AI researchers will simulate human intelligence in a computer, in a robot. At stage two, humans and machines will merge step by step, replacing portions of our brains with mechanical parts. At stage three, AI researchers will reduce existing human intelligence to a pattern of information processing and download this into a computer or a robot. This will lead to an evolutionary advance, actually a leap forward that could lead to cybernetic immortality – that is, immortal intelligent life in a machine that gets constant backups. In traditional vocabulary, we will have removed a person's soul from its native biological body and placed it in a mechanical body.

Note the assumptions. First, what makes us human is our intelligence. Second, intelligence consists of information processing. Third, the transfer of the pattern for information processing from our brain to a machine is feasible. In addition, if we keep the machines in perpetual repair, we can live for ever.[15] We will have achieved cybernetic immortality.

Now, just what corresponding theological assumptions would fit with this technological scenario? It appears that substance dualism fits. Yet, when it comes to theological commitments regarding the human soul, quite a variety of options appear in Christian and non-Christian thinking. So, with transhumanism as a stimulus, how might we plot theological responses?

From Brain to Information Pattern to Soul

The symbol of the soul makes for a powerful concept in the history of Christian theology. Reflecting on what St. Paul says, 2 Corinthians 4:16 "So we do not lose heart. Even though our outer nature is wasting away, our inner nature is being renewed day by day." Martin Luther offers an interpretation. "Man has a twofold nature, a spiritual and a bodily one. According to the spiritual nature, which men refer to as the soul, he is called a spiritual, inner, or new man. According to the bodily nature, which men refer to as flesh, he is called a carnal, out-

[15] Frank J. Tipler, *The Physics of Immortality* (New York: Doubleday, 1994) 305; 309–313. For a theological analysis of Tipler, see Ted Peters, *Science, Theology, and Ethics* (Aldershot UK: Ashgate, 2003) Chapter 17.

ward, or old man."[16] Here we see the soul defined as spiritual and associated with our inner personhood, our essence, our renewal in Christ, our immortality.

What we find in Scripture and interpreters of Scripture such as Luther is symbolic speech, the first level discourse of the Christian faith. When subjected to philosophical explication, we need to ask: what does this mean ontologically or metaphysically? What concept of the soul rises up into the second level discourse of the theologian? In what follows, I will present a spectrum of options, each offering a theological picture of the human soul.

The challenge of transhumanism requires that theologians examine their inventory of soul concepts. Just what do we mean by the soul? And what do we mean by related concepts such as spirit or resurrection? In what follows I would like to draw a spectrum of various understandings of the human soul as found within Christian theology, and in some cases non-Christian theology and even atheism. The dualists are the ones who define the soul in terms of a non-physical or spiritual substance. But the dualists do not stand alone. Many options sit side by side on the spectrum of Christian concepts. Even though substance dualism has garnered some theological credibility in the past, the idea that the soul is a spiritual substance is widely rejected in today's Christian circles as unscriptural and incoherent. Recognizing this makes confronting trans-humanism and cybernetic immortality a bit more difficult, because the theological perspective simply does not match the emerging scientific and technological perspective.

Before we proceed into the spectrum, I would like to flag a distinction between two differing assumptions regarding the soul illustrated by the transhumanism discussion. One is that the soul is the equivalent of the mind. To study the mind is to study the soul, so it is assumed. Yet, theologically, the mind is only part of who we are as a person. In addition to our soul, does our personhood include such things as our body or our spirit? The other assumption is that the word *soul* refers to the essential self. To transfer a soul would be to transfer the essential self. Yet, the matter might be a bit more complicated. Actually, Christian theologians think in terms of a centered self, the unifier of all that is disparate in a person's experience and relationships. Reinhold Niebuhr, for example, reminds us that the soul does not stand alone but rather orients every aspect of our personhood. "The self is soul insofar as it has an experience of the unity. But it is more than soul insofar as

[16] Martin Luther, *Luther's Works*, American Edition, Vols. 1–30, ed. by Jaroslav Pelikan (St. Louis: Concordia Publishing Company, 1955–1967); Vols. 31–55, ed. by Helmut T. Lehmann (Minneapolis: Fortress Press, 1955–1986) 31:344.

it can think of its body as an object even while it is an inner experience of the bodily organic unity."[17] This should alert us to ask just what assumptions regarding the place of the soul in integrated human personhood are being made by each theology of the soul. By identifying these two assumptions, we suggest that theologies of the soul may turn out to be much more complex than what transhumanists assume.

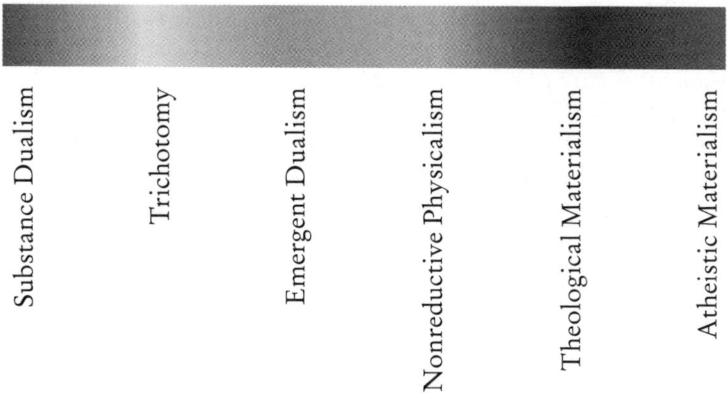

Figure 3: Theologies of the Soul by Martinez Hewlett.
Used with permission.

Substance Dualism

As suggested already, at first glance it appears that transhumanists assume an equivalent to substance dualism, where the information pattern that constitutes the human mind can be extricated from the physical substrate to which it is presently attached. Such dualism has clear precedent in theological tradition.

All Christian theology relies on the word *soul*, to be sure; but what the substance dualist adds is that the soul is a substance, a non-physical or even spiritual substance. The soul is "is an immaterial and spiritual thing," wrote the fourth century Gregory of Nyssa.[18] We human beings are said to be a composite of two substances, one physical and the other spiritual. Even though the spiritual soul resides *with* the body, it is not itself bodily. Like pouring a few drops of motor oil into a bowl of water and watching the colors swirl around, the two

[17] Reinhold Niebuhr, *The Self and the Dramas of History* (New York: Scribner's, 1955) 26.

[18] Gregory of Nyssa, "On the Soul and the Resurrection."

substances are located in the same place but they do not mix. Still further, who we are in essence is found in our soul, not our body. In fact, when we die, our essence will continue on to live while the physical body is entombed in the cemetery. The immortal soul is frequently called the "spiritual soul."

The soul provides the arena within which our mental activities take place. Although plants and animals have souls in a primitive sense of a life force, the highest form of soul, according to the ancient Greeks, is the rational mind. The Greek word from which we get our words 'psychology' and 'psychiatry' *psyche* can be translated as either 'soul' or 'mind'. The chief spiritual goal in the ancient Greek speaking world was to win the struggle between our higher spiritual or mental nature and our lower physical or bodily nature. The appetites of the body tended to distract the mind, to preoccupy the mind with lower things. So the spiritual task became one of self-discipline in which the mind would think higher thoughts and thereby escape the downward pull of the body. This is reflected in the background to the New Testament where St. Paul speaks of the battle between flesh and spirit (Galatians 5).

Christian theologians work extra hours trying to sort out just what is meant by "flesh" and "spirit," because they want to avoid reducing this to a battle between what is physical and what is non-physical. This is because Christians work with a positive view of what is physical. The incarnation blesses what is physical. So, the term "flesh" comes to refer not to the body per se but rather to a condition of the soul, to the soul's self-destructive desires. "We need to ascertain what it is to live after the flesh," says Augustine, "(which is certainly evil, though the nature of flesh is not itself evil)."[19]

Duality, in which the soul differs from the body, seems to fit with our everyday experience of a battle between flesh and spirit.[20] Let me emphasize that it *seems* to fit our experience. What we experience is that we operate with a single centered consciousness. As we think about what we see and hear and smell, there is a single "I" or "me" that is the focus of this thinking. Even when we have a hard time making up our mind and say that "I'm of two minds about this," as soon as we make a decision we become one mind again.

---

[19] Augustine, *City of God*, XIV:2.

[20] "What experience explains my having the belief that I am a soul?" asks Stewart Goetz. "The most plausible answer is that the experience in question is just my inner or introspective awareness of myself as a simple substance that exemplifies psychological properties." "Substance Dualism" in *In Search of the Soul: Four Views of he Mind-Body Problem*, ed. by Joel B. Green/Stuart L. Palmer (Downers Grove IL: InterVarsity Press, 2005) 39.

If we decide to lose weight because it will improve our health, we may undergo a struggle when walking the aisles of the supermarket. We need a strong will to combat the temptation to buy and devour all that appears tasty. It seems reasonable to explain this experience as a battle between flesh and spirit, between our physical bodies and our spiritual souls. Regardless of whether our soul is a separate substance, we experience duality.

Substance dualism within Christianity looks a bit like the ancient Hindu doctrine of the soul. The essential self or *atman* is buried within the body, buried in the darkness of physical strivings, passions, and actions. Bodies and bodily activities are plural, divided, multiple; whereas the soul is unified, one, singular. The body is dark; and the buried soul within us seeks to unbury itself, to rise up toward the light, and unite itself with universal truth. Enlightenment – that is en*light*enment – is the spiritual goal of Hindus. Once we have purified our soul from bodily contamination and attained Enlightenment we will realize that our soul is not finally an individual substance. At this point it is at one with the whole of reality, the All. It is at one with *Brahman*.

While the soul is still contaminated by bodily actions (*karma*), say the Hindus, the soul remains a single spiritual substance with its own independent destiny. It recycles. When we die, our soul proceeds to become reborn in someone else's body. This is the process of metempsychosis or the cycle of rebirth. We know it as reincarnation – that is, from *re* (meaning again) *carnal* (meaning fleshly). Reincarnation means entering the flesh again and again. The final destiny of the soul for Hindus is to jump off the wheel of reincarnation, to leap into the realization of the oneness of *atman* and *Brahman*.

The wheel of rebirth signifies that the soul is eternal. Perhaps more precisely, it comes from eternity and returns to eternity. While visiting our bodies in the temporal realm, the soul becomes an individual substance. The soul is individual, to be sure; but it is individual only as long as it follows the path of incarnation and reincarnation. Each one of us has a soul, but our bodies pull it down into the darkness of physical attention and action to prevent it from liberating itself for a purely spiritual realization. When liberated from its attachment to physical substance, it returns to Brahman and escapes the wheel of rebirth.

When early Christians adopted a version of substance dualism, they insisted on one item that sharply distinguishes their view from Hinduism and from similar versions of metempsychosis among the Greeks. The Christians insisted that the soul did not come from eternity. Rather, the soul began when our bodies began. The substance dualists of today who are Christian insist that God creates a brand new soul for each individual human body. The soul may last forever, but it has a beginning. And, the soul does not get recycled. One incarnation is enough.[21]

Figure 4: Shiva dancing within the Wheel of Rebirth. Picture by Ted Peters.

Climbing Up or Down the Ladder of Being?

Imagine a ladder. On the top of this ladder we locate the fullness of being, the divine essence from which all reality emanates. Below the bottom is non-being, that which does not exist. We will locate the hu-

[21] Early Christians rejected reincarnation for a second reason, namely, in the resurrection it is our former body which becomes transformed and reconnects with the soul. In "On the Soul and the Resurrection" Gregory of Nyssa contrasts what the Christian holds from what the reincarnationist holds this way. "The divergence lies in this: we assert that the same body again as before, composed of the same atoms, is compacted around the soul; they suppose that the soul alights on other bodies."

man soul in the middle, half way between top and bottom. This is the ladder of being we find in Neoplatonism, both a philosophy and a religion during the final centuries of the Roman Empire.

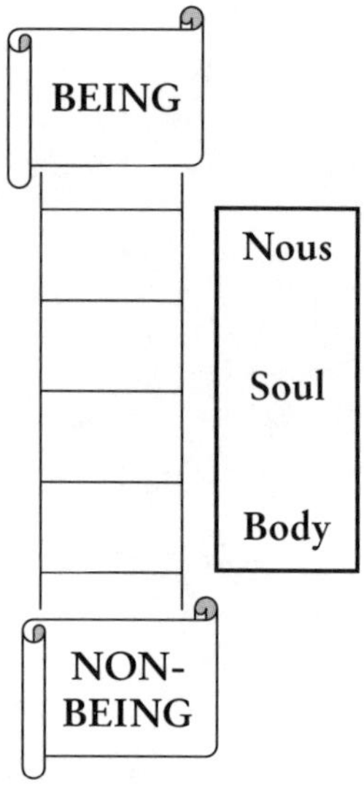

Figure 5: Ladder of Being by Ted Peters.

Between the divine top and the soul on the ladder we find mind, or *nous*. This *nous* is universal; it comprehends intellectually the infinite that is above it. The soul just below has the option of attuning itself to this universal intellect. If the soul attunes itself to thoughts of the infinite, it moves upward in the direction of divinity. However, if the soul turns its attention to what is below it, it focuses attention on what is finite, particular, material, phenomenal, less full of being and hence less real. The downward looking soul constricts itself and contaminates itself with what is physical, what is ephemeral and subject to deterioration and death. Salvation for the soul is to be found in directing itself up the ladder, first to universal *nous* and finally to mystical union with the divine source of all being. What gets saved, according to this Neoplatonic scheme, is the disembodied soul passing through the intellect to realize its divine connection.

The tradition of Augustine, Luther, and Gerhard Forde rejects the idea that we have to climb a ladder to heaven. Rather than ask the soul to climb up the ladder of being, the startling message of the gospel is that God comes down the ladder. In the incarnation in Jesus Christ God descends the ladder right past the soul, down into our finite physical world, and through death even to the abyss of nonbeing. Salvation does not consist of our climbing out of the physical world; rather, it consists of redeeming the physical world.

While interpreting the story of Jacob's vision of the ladder connecting heaven and earth (Genesis 22:18; John 1:51), Luther emphasizes that in Christ the divine nature plummets to the lowest of the low. The divine, "who is the highest and is terrible in His majesty above all creatures, becomes the lowest and most despised ... GOD, who created all things and is above all things, is the highest and the lowest ... Therefore both things are true: the highest divinity is the lowest creature, made the servant of all men, yes, subject to the devil himself. On the other hand, the lowest creature, the humanity or the man, sits at the right hand of the Father and has been made the highest."[22] On the one hand, this amounts to a great reversal. On the other hand, the person of Jesus Christ with the two natures spans the ladder from bottom to top, so to speak.

"The difficulty with the idea of the ladder," writes Forde, "is that it tends to send us off in the wrong direction. It tends to make us concerned with the works of pious sublimation; it involves us in the task of ascending to heaven when we should be seeking like our Lord to come down to earth, to learn what it means to be a Christian here on this earth."[23] Forde was concerned primarily about the relation of the law going up and the gospel coming down the ladder; but it applies in part to the metaphysics of the soul as well.

Although Neoplatonism shares a lot with substance dualism, it's a bit more subtle. The ladder of being places the soul within a dynamic relationship between the universal and the particular, between the material and nonmaterial. Be that as it may, the Christian commitment is to include the particular and the material in the goodness of creation and in the objective of redemption. The Augustinian tradition, in sharp contrast to the Neoplatonic, depicts God coming down the ladder, descending to a level lower than the soul in order to redeem the whole of reality, including body and flesh.

---

[22] Luther, *Luther's Works*, 5:219. Technically, what Luther is explicating here is the *communicatio idiomatum*, "the communion of properties" within the person of Christ. Ibid.

[23] The late and revered Gerhard O. Forde used to rely upon the ladder metaphor to distinguish a meritorious soteriology – "up the down staircase" – in which we climb the ladder to divine acceptance from a "down-to-earth" God of grace. *Where God Meets Man* (Minneapolis: Augsburg, 1972) 8.

Problems with Substance Dualism

So we can see that substance dualism is fraught not merely with philosophical problems, but with theological problems as well. First, as we have just seen, it fails to adequately account for the divine movement of grace that seeks redemption for the whole of reality, the physical dimension included. Now, second, substance dualism is not Scriptural. But, you might retort: doesn't the Bible talk of souls? Oh yes, to be sure. The Bible talks frequently about our soul; and it talks about our spirit as well. It also talks about the Holy Spirit. But nowhere does it describe the soul as a *spiritual substance* especially created by God to make human reasoning possible. So our concern is not whether there is or is not a soul; but rather, whether substance dualism is the best interpretation of what the Bible says.

Genesis 2:7 is one of those passages nearly everyone needs to attend to when dealing with biblical anthropology. Here God formed "a man (*Adam*) from the dust of ground (*adamah*) and breathed into his nostrils the breath of life, and man became a living being (*nephesh*)." This Hebrew word, *nephesh*, referring to our animal life force, is translated "soul" in English. Without soul, we can't live. Who we are as living beings is a whole, a unity of body and soul. Hebrew scholars are reluctant to associate the Old Testament concept of *nephesh* with the thinky, thoughty, immortal intellect of Plato.[24]

When we turn to St. Paul in the New Testament, the soul dies. It perishes (1 Corinthians 15:35–49). Our entire identity drops into nonbeing. St. Paul still believes in resurrection, to be sure. What is raised is our entire person, and the form of resurrection he calls a "spiritual body" (*soma pneumatikon*). Because of this, Christian creeds affirm "resurrection of the body," not immortality of the soul. In summary, the concept of the soul as a spiritual substance that is immortal is not what the Scriptures teach.

A third problem is this: we live with a metaphysical conflict within Christian theology. Despite what we have just said, the idea that the soul is a spiritual substance with immortality does appear in some places in later Christian tradition. The fact is that the tradition includes substance dualism. At no point does it become dogma, yet it is true that over the centuries many Christians have gotten along quite well believing they possess immortal souls. So, we would be reluctant to go

---

[24] "In the OT it never means the immortal soul, but is essentially the life principle, or the living being, or the self as the subject of appetite and emotion, occasionally of volition." N.W. Porteous, "Soul, *The Interpreter's Dictionary of the Bible*, ed. by George Arthur Buttrick (4 Volumes: Nashville: Abingdon Press, 1962) 4:428.

so far as to say that substance dualism and Christian belief are totally incompatible. Substance dualism is not heresy.[25] Well, not quite. The question is this: what is the best understanding of the human soul?

A fourth item to consider is that we as human persons do in fact have the experience of a unity of rational consciousness. We get anxious when on occasion we become flustered by over stimulation and just can't get all our thoughts together. Yet our rational minds seem to have the power to work through the threat of multiple stimuli and organize otherwise scattered thoughts until a unity of consciousness is regained. Our organizing center seems to have a way of exerting itself. Just how do we explain this? Could it be that our rational consciousness is due to the presence of a spiritual substance known as the soul?

This seems plausible to some. This leads one evangelical Christian dualist to write, "Scripture as a whole does not teach that the soul exists. Scripture simply presupposes the existence of the soul because its existence is affirmed by the common sense of ordinary people."[26] Note what this theologian is saying: substance dualism fails the test of being biblical but it passes the test of being experiential. Perhaps British neuroscientist Malcolm Jeeves offers the best alternative here by acknowledging the experience of duality within a higher order personhood that includes unifying consciousness. "There is duality, but not dualism; the ontological reality of 'person' is primary and is neither mental nor physical."[27]

A fifth concern is that so many who take up the body-soul relationship reduce it to the relationship of the brain to the mind. The soul becomes coincident with the human capacity to reason.[28] What the theologian is concerned about is human personhood, and the capacity to reason is only a part of who each of us is as a person. We cannot re-

---

[25] The former Joseph Cardinal Ratzinger, now Pope Benedict XVI, has deliberated carefully on this matter and contends that the "body-soul idea remains unavoidable." Johann Auer/Joseph Ratzinger, *Eschatology: Death and Eternal Life*, Volume 9 of *Dogmatic Theology*, tr. by Michael Waldstein (Washington DC: Catholic University of America Press, 1988) 265. Ratzinger further believes that science buttresses substance dualism. Ibid., 255–256.

[26] Goetz in *In Search of the Soul*, 139.

[27] Jeeves, "Human Nature: An Integrated Picture," in *What About the Soul?*, 30.

[28] The soul is our "intellectual essence," wrote Gregory of Nyssa. This is key to our relationship with God, he thought. "The speculative and critical faculty is the property of the soul's godlike part; for it is by these that we grasp the Deity also." ("On the Soul and the Resurrection") In contemporary neuroscience, some researchers do not limit the soul to reason. They also look for emotions, saying it was Descartes' error to limit the scope of the soul to its rational capacity. See Joseph LeDoux, *The Emotional Brain* (New York: Simon and Schuster, 1996) and Antonio R. Damasio, *Descartes' Error: Emotion, Reason, and the Human Brain* (New York: Avon, 1994).

duce who we are to this one aspect. The problem here is not necessarily the association of the soul with reasoning; rather, it's the relationship of the capacity to reason with all the rest that makes us the persons we are. An integrated anthropology would include our reasoning soul but not define our individual essence according to our soul alone.

Still one more distinctively theological problem arises. To identify the soul with any innate human capacity whatsoever is a theological problem. Whether identifying the soul with reasoning or any other human ability, such an anthropology risks obscuring an extremely important Christian principle, namely, God relates to us graciously. God loves us before we can love ourselves let alone love God in return. God approaches us, calling us into being and into relationship with God. God's call to us is not dependent on some capacity we have, nor even on a capacity God would give us along the way. Salvation requires a divine action; it does not come as an extension of a property we human beings possess. "Nothing in the created human being is intrinsically immortal," writes Joel Green. "Resurrection and embodied afterlife are God's doing, divine gift."[29]

If in common parlance we would like the word 'soul' to connect with what we believe is the essence or core of a person, this may be okay. However, theologically, soul is a much more relational concept. It has to do with our relationship to God. Our essence is determined by our relationship to the God who calls us into being and with whom we will have an everlasting sharing. The human soul is not an independent immortal substance. Rather, what we might call 'immortality' or life beyond death will be due to a creative act on the part of the God who loves us. God will act to raise us from the dead to live eternally within the divine life. Our future resurrection is in no way related to possessing now a spiritual substance that is immune to death. Instead, our resurrection is totally dependent on a future act of God, the act whereby God will fulfill the Bible's promise to raise us into the new creation.

Materialism, Atheistic and Theistic

When philosophers of religion get their intellectual fingers wrapped around an issue such as the human soul, they squeeze out more distinctions than Minute Maid can squeeze out orange juice. This leads to a spectrum of views of the soul with substance dualism at one end and the abject denial of the soul by atheistic materialists at the other

[29] Joel B. Green, "Resurrection of the body; New Testament Voices Concerning Personal Continuity and the Afterlife," in *What About the Soul?*, 98.

end. Strewn in between are variants. Here we'll identify some of these. Then we'll turn to saying more about what we mean by a relational and integrated understanding of the human person inclusive of body, soul, and spirit.

If we place substance dualism at one extreme end of a soul-talk spectrum, at the other extreme we place eliminative materialism, especially *atheistic materialism*, which dubs soul talk nonsense. Materialists answer to a number of different names such as naturalists, humanists, reductionists. The only reality is physical reality, say eliminative materialists. No such thing as a spiritual substance exists. When it comes to what we experience as a unity of consciousness or even rational thinking, materialists are confident that with enough research we will eventually be able to demonstrate that biological processes in the human brain exhaustively explain what happens in the human self or psyche. What we think of as soul activity can be reduced to biological activity. And human biology comes to us from the long history of evolution, not from God. So goes the argument of atheistic materialists.

Harvard philosopher Daniel C. Dennett calls himself a 'naturalist', advancing a materialist view when it comes to the soul. He rejects substance dualism as pre-modern, as having been discredited by modern science. "This idea of immaterial souls, capable of defying the laws of physics, has outlived its credibility thanks to the advance of the natural sciences."[30] The mental processes of the mind can be reduced to the physical processes of the brain. "Our minds are just what our brains non-miraculously do, and the talents of our brains had to evolve like every other marvel of nature."[31]

If we were Buddhists, we might want to place ourselves in the camp with the atheistic materialists. Metaphysically, Buddhists deny that there is such a thing as a human self, that what we experience as our individual soul is a delusion. The religious task is to face the ultimate reality; and the ultimate reality is that we have no soul, no self (*annata*). What appears to be the person we think we are is actually only a heap of physical processes, physical processes that are finally hollow and empty of an inner essence. Facing realistically the emptiness of inner essence would be a religious achievement for a Buddhist. Once we've faced this reality, then we become released from the nexus of physical interactions; and what we took to be our inner self dissolves into the totality of reality. We experience nirvana.

Curiously, however, within Buddhism we also find the doctrine of reincarnation. On the one hand, Buddhists inherited from early Hin-

---

[30] Daniel C. Dennett, *Freedom Evolves* (New York: Viking, 2003) 1.

[31] Ibid., xi.

dus the concept of rebirth even though, on the other hand, they deny that such a thing as a spiritual substance transfers itself from one incarnation to the next. What we end up with is a loosely fitting combination of substance dualism and atheistic materialism.

Another variant still within the materialist camp is Christian *theological materialism.* Christian materialists find themselves quite at home with modern science. They are at home with biological reductionism, and quite at home with denying the existence of a separate spiritual substance. Although Christian materialists can give up on the idea of a separate spiritual substance, they don't give up on resurrection. They strongly emphasize that both Bible and tradition – tradition in the form of the Apostles' and Nicene creeds – explicitly and unambiguously teach resurrection of the body, not immortality of the soul. "Christian faith is not about ensoulment, but of embodiment," writes Jürgen Moltmann. The word 'body' refers to "the experienced and lived body, with which I am directly and spontaneously identical: I am body = This body is me."[32]

This materialist emphasis implies that death is total. Kevin Corcoran, who calls himself a "constitutionalist" because it is the body that constitutes the person, writes, "for reasons having mostly to do with biblical interpretation, I am of the opinion that death marks a real and tragic end to our existence ... The Christian doctrine is *resurrection* of the body."[33] No such thing as immortality is possible, say the materialists, unless God takes whatever action is necessary to raise the dead. If and when God takes such action, it will consist of raising the body; and the continuity of our identity in eternity will include bodily identity.

Emergent Dualism and Nonreductive Physicalism

Lying between substance dualism on one end and materialism in its various forms on the other end, in between we find two more options, emergent dualism and nonreductive physicalism.

According to emergent dualism, we may still speak of human beings as possessing a soul. What distinguishes this position is where the soul comes from. Whereas for substance dualists the soul comes from

---

[32] Jürgen Moltmann/Elisabeth Moltmann-Wendel, "To Believe with All Your Senses: The Resurrection of the Body," The Catholic Theological Society of America: *Proceedings of the Sixtieth Annual Convention, June 9–12, 2005,* ed. by Richard C. Sparks, Vol. 60 (Berkeley: Newman Hall, 2005) 2.

[33] Kevin Corcoran, "The Constitution View of Persons," in *In Search of the Soul*, 168, Corcoran's italics.

a special and individual act of God's creation, for emergentists the human soul emerges for the human race through evolution and emerges in the individual through fetal development. First comes the body, then comes the soul. The soul is an emergent property. The core meaning of 'emergence' is that when material elements of a certain sort are assembled in an optimum way, something new and more complex comes into being, something that had not existed before. Some refer to this as 'holism' which indicates that new wholes are greater than the sum of the parts that made them up. Hence, when all the chemicals that make up the human body come together in an optimum way, a new whole – that is, a person – emerges. More, a soul emerges. This soul is distinguishable from the body and brain that had been its source. "The fundamental idea is actually rather simple," writes William Hasker. "As a consequence of a certain configuration and function of the brain and nervous system, a new entity comes into being – namely, the mind or soul."[34]

Keith Ward writes, the soul "is a point of subjectivity and transcendence, of rational understanding and responsible action, which comes to be at a particular stage of the emergent interactions of spatial, material substances. Once it is generated, it continues to have a place in those physical interactions, to respond to them and realize itself in them, as their actuating principle, as Aquinas put it."[35] Once the soul has emerged in evolutionary history, it opens the human psyche to transcendence. "Thus, though it is truly material in origin, there is a very real sense in which the soul looks beyond the material world for its proper fulfillment. It looks to a presence and perfection and purpose beyond all material form – to the reality of God, its true source and goal."[36]

Ward brings us to a close sister of emergence, to nonreductive physicalism. Nonreductive physicalism accepts the idea of emergence in emergent dualism, but it rejects the independent substantial status for the emergent soul or mind. *Nonreductive physicalism* affirms that (a) it is unnecessary to postulate a second, metaphysical entity, such as a soul or spirit, to account for human capacities (or even what we have been calling the experience of duality), and that (b) human behavior cannot be explained exhaustively with recourse to genetics or neuroscience.[37] The metaphysical status of the soul here is closer to what the materialists affirm. No soul exists, though we certainly experience our minds.

---

[34] William Hasker, "On Behalf of Emergent Dualism," in *In Search of the Soul*, 78.

[35] Keith Ward, *Defending the Soul* (Oxford: One World Publications, 1992) 148.

[36] Ibid., 149.

[37] Warren S. Brown, "Conclusion: Reconciling Scientific and Biblical Portraits of Human Nature," in *Whatever Happened to the Soul? Scientific and Theological Portraits of Human Nature*, ed. by Warren S. Brow/Nancey Murphy/H. Newton Malony (Minneapolis: Fortress Press, 1998) 213–228.

The human person is constitutionally physical, but not reducible to the physical. This position can be contrasted with eliminative materialism, which holds that mental entities do not exist at all. It can also be contrasted with reductionist materialism, which holds that mental entities exist but can be reduced to physical events. Nonreductive physicalists reject substance dualism in favor of materialism, but the human mind is an emergent that is not reducible to its physical base.

The mind is a coordinated product of brain activity, but it is more than mere brain activity. Nancey Murphy puts it this way: "The nonreductive physicalist says ... that if there is no soul, then these human higher capacities must be explained in a different manner. *In part*, they are explainable as brain functions, but their full explanation requires attention to human social relations, to cultural factors, and most importantly, to God's action in our lives."[38]

Like the theological materialists, Christian nonreductive physicalists affirm resurrection of the body. They emphasize that immortality comes not from the immortal quality of a spiritual substance; rather, it comes solely as an act of God whereby we will be raised. Some sort of physical identity as well as identity of internal consciousness must be carried into the resurrected life if it is we, and not someone else, who enjoys resurrection. What Murphy adds that makes her position especially interesting, is that the moral character that has developed in this life is carried through by God into the next. "I propose that identity of persons depends as much on character identity as it does on memory/consciousness and bodily continuity."[39]

Resurrection and the Human Soul

Creedal Christians affirm "resurrection of the body." The words "immortal soul" or "immortal spirit" do not appear in the creeds, despite the observation that many Christians find comfort in such a belief. The fact remains that it was the Easter resurrection of Jesus in the body that provided the shock therapy in messianic thinking that led to affirming resurrection of our body. As cited in an earlier chapter, Adolph von Harnack, wrote, "the resurrection of Jesus became the sure pledge of the resurrection of all believers, that is of their real personal resurrection. No one at the beginning thought of a mere immortality of the spirit."[40]

---

[38] Nancey Murphy, "Nonreductive Physicalism," in *In Search of the Soul*, 116, Murphy's italics.

[39] Ibid., 186.

[40] Adolph Harnack, *History of Dogma*, 7 Volumes (New York: Dover, ³1900) 1:85.

This affirmation of bodily resurrection does not in itself settle the dispute regarding the best theory of the soul, however. Emergentists or nonreductive physicalists do not gain an edge because they come close to materialism. Even Gregory of Nyssa, who held a virtual substance dualism, still believed that the disembodied soul would be reunited with the original body on the day of general resurrection. "When our race has completed the ordered chain of its existence as the ages lapse through their complete circle, this current streaming onward as generation succeeds generation will cease altogether, but also that then, when the completed Universe no longer admits of further increase, all the souls in their entire number will come back out of their invisible and scattered condition into tangibility and light, the identical atoms (belonging to each soul) reassembling together in the same order as before."[41] This is the way a fourth century Christian theologian writing in Greek thought. He thought all the original atoms of our body would be reassembled. This ancient church father combines substance dualism with resurrection of the body.

For Latin speaking St. Augustine, the resurrection would mark a suffusion of the body with the spirit. "Not that it [the body] is converted into spirit, as some fancy from the words, 'It is sown in corruption, it is raised in incorruption' [1 Cor. 15:42], but because it is subject of the spirit with a perfect and marvelous readiness of obedience, and responds in all things to the will that has entered on immortality – all reluctance, all corruption, and all slowness being removed."[42] The reassembled body will be a spiritually transformed and redeemed body.

Approximately six centuries later, St. Thomas Aquinas thought he could affirm both a present soul and a future resurrection. Following Aristotle, Thomas taught that body and soul belong together as a unity. The soul is the form of the body. Yet, as a spiritual substance, the soul could separate from the body.[43] The soul has a single eternal nature, and this one nature exists in one mode in the body and a different mode outside the body. When separated at death, it continues. The soul reunites with the body at the future resurrection.

Most contemporary theologians drop the substance dualism and emphasize more the transformation of the body. Important here is the "spiritual body" spoken of by Paul in 1 Corinthians 15:42–44. New Testament scholar Joel B. Green connects personal transformation with cosmic transformation, with new creation. "The capacity for 'afterlife' is not a property of humanity, but is a divine gift, divinely enacted …

---

[41] Gregory of Nyssa, "On the Soul and the Resurrection".
[42] Augustine, *City of God,* XIII:20.
[43] Thomas Aquinas, *Summa Theologica*, I.I.Q. 89, Art. 1.

in eschatological salvation, we are not rescued from the cosmos in resurrection, but transformed with it in new creation."[44]

What we see here in the fundamental Christian vision is acknowledgement of duality, but not necessarily dualism. We also see affirmation of bodily transformation, With or without a distinctively nonmaterial soul, bodily resurrection is inextricable to personhood in Christian anthropology.

Body, Soul, and Spirit as a Trichotomy

The reaction against Cartesian substance dualism in both culture and theology seems to set one agenda, namely, how do body and soul relate? Yet, the Christian symbol system is not strictly dichotomous. It does not divide the human person into two parts. More frequently, it is trichotomous, integrating body and soul with spirit.

A position widely ignored by philosophers of religion yet quite important on our spectrum of options is *trichotomony.* The anthropology of trichotomists views the human person as a whole made up of body, soul, and spirit. The body is physical, quite obviously. But, soul and spirit are not conflated into a 'spiritual soul' as in substance dualism. According to trichotomists, it is acceptable to identify the soul with the mind, with the individual mind. The spirit, on the other hand, makes possible a supra-physical unity or bond one individual can share with others, with the church, and with God. Each of us enters the world at birth with both a human soul and a human spirit. In baptism the human spirit is replaced by God's Holy Spirit, at least according to some of the earliest Christian trichotomists in post-biblical times. It is in the domain of spirit where the overlap between the human life and the divine life takes place.

One thing trichotomy has going for it is that it is biblical. In moments of emphasis the New Testament refers to body, soul, and spirit (1 Thessalonians 5:23). Without specifying precise definitions for each, the terms body, soul, and spirit together constitute a whole human person, the beloved of God.

Again, no metaphysical commitment to separate substances – a physical substance, a soul substance, and a spiritual substance – is made in the Bible. Perhaps some of the early trichotomists presumed that we are dealing with three substances; but this certainly has not gained a following in Christian tradition. One could certainly hold to affirming we need all three to be a person – body, soul, and spirit – with-

[44] Green, "Resurrection of the Body" in *What About the Soul?*, 100.

out making any metaphysical commitment to multiple substances. A non-substantialist appropriation of the scriptural picture of a person as body, soul, and spirit should appear on the agenda of Christian anthropology.

Conclusion: Relationality and Integrity

Before the theologian cooks up a tasty metaphysics of the soul, I would like to suggest that the dish include two contrasting yet complementary ingredients, relationality and integrity. These two roughly correlate with what Paul Tillich described as a pair of ontological polarities, participation and individuation.[45] We find ourselves slung between being and nonbeing, between the dynamism of interaction and the centering into ourselves as an individual. The phenomenon of being human includes the integration of our relationships into our personhood.

The substance dualism which dominated Western culture since the Enlightenment is becoming effete in general, not just in theology. We are understanding more clearly what has probably been true all along, namely, reality is itself relational in structure. Acknowledging this is requisite for theology today, insists Paul Sponheim. "The notion of relationship figures more prominently in how we look at reality. Perhaps it does so because we 'really' are created in and for relationship, and we are catching on to the gift and task of relationship in our living."[46]

Can we apply relationally to the soul? I believe we can, and ought to. The soul is not an independent entity, let alone an independent spiritual substance. Rather, the term *soul* is a symbolic placeholder, so to speak, to identify that dimension of who we are that connects us with God. If Jesus identifies the soul with our eternal destiny, and if God and only God's action is responsible for that eternal destiny, then the case for the soul as inclusive of our relationship to God seems to be made. Theologically, this would be the case regardless of the metaphysics that describe how it is the case.

Augustine believed the soul requires a relationship with God for it to have life or to sponsor life for the body. The life of the soul is dependent on its relationship to God, the source of life. "Although the human soul is truly affirmed to be immortal … the soul is justly said

---

[45] Paul Tillich, *Systematic Theology* (3 Volumes: Chicago: University of Chicago Press, 1951–1963) I: 174–178.

[46] Paul Sponheim, "Closer to the Shifting Ground: the Rise of Relationship in God-Talk," *Dialog* 44: 3 (Fall 2005) 231.

to die … The soul, then, lives by God when it lives well, for it cannot live well unless by God working in it what is good."[47]

This understanding of the soul would need a corresponding understanding of God's relationship to the world, or at least to God's relationship to the work of salvation. We find this in many places in contemporary theology, one of which is open theism. Clark H. Pinnock gives articulation to this position: "*Open theism is a relational and Trinitarian doctrine with an emphasis on God as personal and interactive.*"[48] No one "can destroy both soul and body in hell" if the relationship we have with our eternal God remains sound.

The other ingredient is integrity. To have integrity is to be integrated around a center. By integrity here I refer to the integrity of the person, of which the soul is an integral part. The soul is integrated into our centered self, into our personhood. The soul – at least the truncated view of the soul as strictly our reasoning function – is not our personhood in itself. Even if posthumanists are able to up-load our entire mental processes into a new computer and perpetuate what we think, they would not be able to take our entire personhood with them. We are more than our minds, more than our brains, more than our souls. Malcolm Jeeves makes this point forcefully. "We are persons – not brains, not minds, not bodies, but persons … A holistic approach to understanding human nature focuses on the person."[49]

Christian anthropology has an investment in the viability of the person-in-relationship model of the human reality. With the advance of the cognitive sciences and posthumanism, this model is under double threat. From one direction, the threat is that our personhood could be reduced to our minds, and our minds reduced to brain activity. This is the threat of reductionism. From the other direction, posthumanism, which also reduces our personhood to our minds, threatens to detach our minds from our bodies and disaggregate who we are as whole persons.

Intellectual honesty obligates the theologian to ask whether prognostications from either direction are accurate. Do these sciences and technologies have a purchase on reality? If so, what might the implications be for Christian anthropology?

It appears that neither has credibility. The cognitive sciences work with the hypothesis that the mind can be entirely mapped on the brain, but it is by no means proven. It's an assumption. To date, the jury is still out. With regard to post-humanism, critics seem to believe that

---

[47] Augustine, *City of God*, XIII:2.

[48] Clark H. Pinnock, "Open Theism: An Answer to My Critics," *Dialog* 44:3 (Fall 2005) 237, Pinnock's italics.

[49] Jeeves, "Human Nature: An Integrated Picture," *What About the Soul?*, 174.

this is a pipe dream that will eventually go the way of all pipe smoke, wafting away.

Be that as it may, Christian theologians need to remain vigilant regarding our biblical exegesis and attention to common human experience in the face of new sciences. Our understanding of ourselves as biological beings and as children of God is likely to grow through such interchanges, even if to date it is less than clear exactly how.

Chapter 8

Science in Pastoral Ministry

The charge of the parish pastor or priest is threefold: *Preaching, Teaching*, and *Counseling*. Of course, there is more to the weekly ministry than these three alone. Preaching takes place within a planned worship setting with liturgy and the celebration of the sacraments. Teaching involves administering a Christian education program and overseeing Bible studies. And counseling is only the core of conversation taking place during hospital visitation, home visitation, and numerous other opportunities for inter-personal engagement. A faithful pastor finds precious little time to stop, read, think, meditate, pray, and discern. Yet, this is the inescapable lot for those who feel they have been called by the Holy Spirit to ordained service.[1]

The worldwide dialogue taking place between natural scientists and Christian theologians as well as intellectual leaders in many of our world's religions may appear to the parish pastor like an ivory tower luxury. Dialogue for the purpose of exploring ideas, searching for overlaps and connections between one's faith in God and the scientific understanding of the natural world, look to the busy pastor like a leisure time activity. It looks like armchair speculation. No parish pastor beset with the burdens of an already impossible schedule, it seems, could justify taking the time to indulge in such a speculative enterprise.

Yet, the revolutionary new rapprochement between science and theology provides the parish pastor with a treasure chest of intellectual jewels that could enrich his or her preaching, teaching, and perhaps even counseling ministries. To leave this treasure chest unopened would be to deny oneself a wealth of resources for parish ministry. In what follows, we will identify some areas of the new dialogue between science and theology that could definitely enhance the effectiveness of preaching and teaching, if not counseling and pastoral care as well.

In this chapter we will first look at pastoral hermeneutics, which includes the prophetic and constructive tasks of biblical preaching. We will then review eight different models of understanding the relationship between science and faith, half that draw upon the image of

[1] This chapter is adapted from Ted Peters, "Contributions from Practical Theology and Ethics," for the *Oxford Handbook of Religion and Science*, ed. by Philip Clayton.

warfare and half which advocate peaceful cooperation. We will give special attention to the controversy surrounding Darwinian evolution. Finally, we will turn to the frontier of genetic research and the ethical issues surrounding the phrase "playing God," recommending that the pastor demythologize science while appropriating science to a theological understanding of the world in which we live.

Pastoral Hermeneutics

Why might interest in modern science be of value to the parish preacher or teacher or counselor? After all, the task of a church leader is to interpret the Bible, not natural science. Yet, the interpretation of the Holy Scriptures on behalf of listening ears in the context of the modern world requires attention to the modern worldview, which has been determined primarily and extensively by natural science.

Yes, interpreting the Bible is what distinguishes the work of the Christian parish pastor. Yet, the dialogue with natural science can aid the pastor in this task of interpretation, first and foremost in preaching. The preacher will need to deal with science on two fronts: first critically as a myth to be transcended by the proclamation of the gospel; and, second, as an essential component in post-critical worldview construction.

To proclaim the Word of God in the sermon in such a manner that it becomes the Word of *God* and not merely that of the preacher is the unenviable charge to the conscientious pastor. "When the Gospel is preached, God speaks," wrote Karl Barth.[2] The presence of the Word of God makes the pulpit a sacred and awesome place. In order to help insure that the Word of God comes through with impact, the pastor in the pulpit needs to attend to two contradictory or paradoxical thrusts within the sermon's structure, worldview-deconstruction and worldview-construction or reconstruction. The first is prophetic; the second is pastoral. The first is critical; the second is post-critical. Let us start by saying a few things about the critical task of worldview-deconstruction.

God is not of this world. God transcends this world. This does not mean that God is distant. It means that God cannot be controlled by anything we do or anything we think. God cannot be subordinated to anything we believe or imagine. God comes to us from outside our imaginations, challenging and cracking through our previously accepted

[2] Karl Barth, *The Preaching of the Gospel.* (Louisville: Westminster John Knox Press 1963) 12.

understandings of ourselves and our world. To proclaim the gospel as God's Word, the preacher needs to challenge our this-worldly assumptions and loyalties so that we can open ourselves to a message that comes from beyond.

Cultures and individuals construct in their imaginations worldviews or pictures of reality that include the whole world and themselves in it. Such views of reality provide the conceptual framework within which we understand our place in an otherwise immense and mysterious cosmos. Yet the God who revealed himself to the people of ancient Israel is simply not of this world. The God of the first commandment – make no images! – transcends all of our images of God. To proclaim the Word of God includes smashing our understanding of how the world works in order to reveal a higher reality, God. The preacher by example and by message needs to communicate to his or her listeners: don't trust what you imagine the world to be like; trust God instead.

During the central third of the twentieth century when neo-orthodox and existentialist theology reigned, most theologians referred to this as preaching the "kerygma" or preaching the gospel, accompanied by "demythologizing." *Demythologizing* was (and still is) an important term. It did not mean eliminating the Bible's mythological worldview and replacing it with a scientific worldview. Rather, it meant interpreting the message of Jesus Christ distinguishable from any worldview. Rudolf Bultmann wrote, "Its [*de-mythologizing*] aim is not to eliminate the mythological statements but to interpret them. It is a method of hermeneutics."[3]

For this discussion, the word 'myth' is nearly equivalent to 'worldview' or 'world picture'. The prophetic task or critical task of the preacher is to break through all worldviews in order for the kerygma to call listeners to personal trust in God, trust in the God who transcends this world and transcends our imagination of how this world works. Preaching consists of a call to faith, with faith understood as existential trust in a God who remains mysterious and unknowable, even if trustworthy. By "world" we mean our worldview or world of meaning, *Weltanschauung*. This needs to be destroyed, or in postmodern terms "deconstructed," so that our attention can be drawn toward the divine grace that comes to us from beyond the world. According to this prophetic understanding, the preacher enters the pulpit ready to smash idols – that is, to critically dismember idols in the form of ideas or concepts or beliefs about how the world works. In radical obedience to the First Commandment against graven images, the preacher

[3] Rudolph,Bultmann, *Jesus Christ and Mythology* (New York: Charles Scribner's Sons, 1958) 18. See: David Ferguson, "Entmythologisierung," RGG[4], 2:1328–1330.

proclaims that God comes from beyond all our mental images of this world and confronts us with the decision to either have faith in the transcendent or continue to rely idolatrously upon our this-worldly imagination.

World-destruction is the prophetic task of the preacher. What I here call the *prophetic* task is built on Paul Tillich's "Protestant principle," which "emphasizes the infinite distance between God and man."[4] By criticizing, challenging, and deconstructing the worldview within which we live, the preacher calls us to faith in the infinitely transcendent God who has revealed the Godself as gracious in the cross of Jesus Christ. The limit to this prophetic preaching task is that it can not on its own tell us how to live in this world, even if it is a world partially constructed by the human imagination. Human beings cannot live on a day to day basis on prophetic preaching alone. Deconstruction is not enough. We all need to envision ourselves within a world that is packed with meaning, existential meaning as well as conceptual sense. This leads to the second of the preacher's tasks, namely, to construct a worldview where all things are oriented around the God of grace. This is the pastoral element in the sermon.

Since the days of St. Thomas Aquinas, the job of the systematic theologian has been to describe all things in the world in light of their relationship to God. This is the job of the contemporary preacher (and teacher) as well as the theologian. The meaningfulness of the life of every person sitting in a church pew is conditioned by the picture of the world that the preacher paints. The preacher is charged with the task of painting a picture of reality that orients all things to the God who transcends all things. This is a most difficult task, to be sure; yet nothing less is demanded.

Once the preacher has called us out of the world to listen to the eternal kerygma or Word of God, then the preacher turns around and calls us back into the world with a new orientation. Our response to the gospel re-orients our understanding of ourselves and of the world within which we live. This happens, whether the preacher is aware of it or not. It is better to be aware.

Implicit in the sentences and metaphors and images and allusions the preacher employs in the sermon is a description of a single comprehensive reality in which both the listener and God are components. This is as important as it is inescapable. The verbal pictures of the world the preacher draws are the primary vehicle for evoking a sense of meaning, of belonging, of orientation, of welcome, of acceptance.

---

[4] Paul Tillich, *Theology of Culture*, ed. by Robert C. Kimball (London/New York: Oxford University Press, 1959) 68.

Hermeneutical philosophers such as Hans Georg Gadamer remind us of the constant dialectic of what is said and what is unsaid.[5] Whatever we say, *das Gesagte*, is always accompanied by a context of what is unsaid, *das Ungesagte.* No matter what the focus of the sermon, the frame is co-present. No matter what act takes center stage, the backdrop is co-present. No matter what the theme, the context is co-present. The comprehensive picture of reality or worldview of the preacher is inescapably co-present with every emphasis or point or theme. What the conscientious preacher needs to do is give reflective attention to the horizon of the unsaid, to the worldview that he or she wants to accompany the kerygma.

Language and worldview belong together. Tom Rogers makes this point. "The gospel does bring change and resulting *challenge*, yet the way we language that *challenge* from the pulpit determines a great deal. The language choices preachers make influence the theological understandings with which hearers leave."[6]

Our concern here is this: what role does modern science play in structuring the worldview and the language that frames the sermon? Note what is not being said here. Our concern is not to demand that modern science provides the content of the sermon. The kerygma or gospel provides the central content. The role that science plays is found in the framing worldview, whether to be prophetically demythologized or to become a component in worldview reconstruction.

Science is an unavoidable factor in describing persuasively to virtually all modern people just how our world works. It is with this in mind that Philip Clayton writes while discussing physical causality, "Remember that the question is not how to *prove* that God is active in the world at particular moments, but rather how to think this possibility in a manner that does not conflict with what we now know of the world."[7]

Science is the reigning myth, so to speak. Science sets the standards for credibility. No one can live with a sense of truthfulness in a worldview that is irreconcilable with the picture of nature drawn by physicists, chemists, biologists, and ecologists. Whatever the preacher assumes at the level of the unsaid will become implicitly audible to his or her listeners; and if it is irreconcilable with science the credibility of everything the preacher says will be unnecessarily doubted.

---

[5] Hans-Georg Gadamer, *Truth and Method*, tr. by Joel Weinsheimer/Donald G. Marshall (New York: Continuum, 1994) 469.

[6] Thomas G. Rogers, "*What* and *How* of North American Lutheran Preaching" *Dialog* 43:4: 263–271 (2004) 270, Rogers' italics.

[7] Philip Clayton, *God and Contemporary Science* (Grand Rapids: Wm. B. Eerdmans, 1997) 193, Clayton's italics.

The Scientific Worldview in Preaching and Teaching

Just how might the parish pastor include natural science in the worldview he or she seeks to construct? This larger question is prompted by a set of smaller questions: do we want to perpetuate the widespread belief that science and faith are at war with one another? Do we want people of faith to fear the sciences as threats to religious belief? Do we want members of congregations to suspect that the scientists who are also members may be secretly committed to atheism? Do we want our young people considering a future career to eschew studying science out of the fear that it may be of the Devil?

To address these questions directly or indirectly in preaching and teaching will require on the part of the parish pastor or priest at least a modicum of sophistication regarding just how science and religion in fact relate to one another. Surprisingly, multiple ways of construing the relationship between science and faith make up the currency of today's idea exchange. Here we will look at eight models or patterns of interaction between science and faith, suggesting directions for pastoral assessment and employment.[8]

The First of Four Warfare Models: Scientism

The first four models fit the widespread belief – myth, if you will – that science and faith are at war. Note the word 'model' here. A model is a conceptual image. It provides a filter for interpreting what is going on. It does not make any absolute claims. Because we have multiple models, no one model is considered the exclusive right one. The parish pastor has no investment in selecting the right one to tout or promote; rather, the pastor in the pulpit needs to employ the various models for prophetic and constructive purposes.

The first model is *Scientism*. In the contemporary West, the term *scientism* refers to naturalism, reductionism, or secular humanism – that is, the belief that there exists only one reality, namely, the material world. Further, science provides the only trustworthy method for gaining knowledge about this material reality. Science has an exhaustive monopoly on knowledge. It judges all claims by religion to have knowledge of supernatural realities as fictions, as pseudo-knowledge. All explanations are reducible to secularized material explanations.

---

[8] Ted Peters, *Science, Theology, and Ethics* (Aldershot UK: Ashgate, 2003) chapter One; Ted Peters, "Science and Religion," *Encyclopedia of Religion*, ed. by Lindsay Jones (14 Volumes: New York: MacMillan, Gale, ²2005) 12:8180–8192; and Ted Peters, "Naturwissenschaft und Religion: Ein wachsender Forschungsbereich," *Verkündigung und Forschung* 49:2 (2004) 52–71.

According to this model, religion loses the war by being declared false knowledge.

Sociobiologist Susan Blackmore provides a minimal case for the superiority of science over religion. "Science is, in some sense, superior to religion ... At the heart of science lies the method of demanding tests of any idea. Scientists must predict what will happen if a particular theory is valid and then find out if it is or not ... This is not what religions do. Religions build theories about the world and then prevent them from being tested ... I do defend the idea that science, at its heart, is more truthful than religion."[9]

Peter Atkins, SmithKline Beecham Fellow and tutor in Physical Chemistry at Oxford, shows a bit more aggression. "My conclusion is stark and uncompromising. Religion is the antithesis of science; science is competent to illuminate all the deep questions of existence, and does so in a manner that makes full use of, and respects the human intellect. I see neither need nor sign of any future reconciliation."[10]

Harvard biologist Richard Lewontin similarly avers that the challenge is to get the public "to reject irrational and supernatural explanations of the world, the demons that exist only in their imaginations, and to accept a social and intellectual apparatus, *Science*, as the only begetter of truth ... Materialism is absolute, for we cannot allow a Divine Foot in the door."[11]

Oxford's Richard Dawkins, author of *The Selfish Gene*,[12] would like to increase the spoils of war for atheists. He wants to fight for some of the charitable contributions that people give to religious causes. In his acceptance speech at receiving the "In Praise of Reason" award in 1992, he said, "Don't let us allow religion to walk away with the awe factor. Science has orders of magnitude more to offer in this field ... And, to become a bit mercenary about it, there's that 54 percent of the charitable donations."[13]

Coming from the other side of the battlefield in the same war, Rustum Roy, Evan Pugh Professor of Geochemistry at Pennyslvania State University, declares: "*Scientism must be destroyed.* By *scientism* I mean the absurdly reductionist belief that all truth can be learned and all reality described through science (never defined) and only through

[9] Susan Blackmore, *The Meme Machine* (Oxford/New York: Oxford University Press, 1999) 202–203.

[10] Peter Atkins, "Religion the Antithesis of Science," *Chemistry and Industry*, (20 January 1997), comment section.

[11] Richard Lewontin, Review of *The Demon Haunted World* by Carl Sagan in *New York Review of Books* (January 9, 1997).

[12] Richard Dawkins, *The Selfish Gene* (New York: Oxford University Press, ²1989).

[13] Richard Dawkins, "The Awe Factor," *Skeptical Inquirer*, 17 (Spring 1993) 243.

science. 'Only' is what distinguishes scientism from what all of us in mainstream science believe about science."[14]

Should the preacher take sides in this war? Yes, indeed.

Against scientism, the preacher needs to speak critically and prophetically. The world of scientism is a sealed off natural world that has closed its doors and windows to transcendence. Not only is it anti-religious, it only pretends to be scientific. Actual science as a research enterprise does not need the ideology of scientism. One can enjoy perfectly good science without adding this ideology of materialism or reductionism. The preacher can help listeners to distinguish between healthy science and unhealthy scientism.

The Second Warfare Model: Scientific Imperialism

The second model on our list is *scientific imperialism*, a close ally of scientism. Scientific imperialism does not outrightly dismiss religion. Rather, it uses materialist reductionism to explain religious experience and reassess theological claims. Scientific imperialists grant value to religion and religious contributions to society. They may even grant the existence of God. Yet, scientific imperialists claim that science provides a method for discerning religious truth that is superior to that of traditional theology. Physicist Paul Davies, for example, says that "science has actually advanced to the point where what were formerly religious questions can be seriously tackled [by] the new physics."[15] Frank Tipler goes even further than Davies. Tipler believes he can provide a physical explanation for immortality, providing a scientific eschatology superior to that of traditional religion. He then declares that in the future the study of theology will take place within departments of physics.[16] In these two cases, previous religious discoveries of God and immortality are affirmed; but science is now said to provide a better explanation than outdated theology can.

Sociobiologist E. O. Wilson proffers a biological explanation for cultural evolution including religion and ethics.[17] The function of religion is to provide cultural support for reproductive fitness and the perpetuation of specific genetic codes. Religion obeys epigenetic rules – that is, religion serves the genes. From now on, sociobiologists will provide us with a scientific account of religion that is more accurate than what a theologian can provide. "Science for its part will test relentless every

---

[14] Rustum Roy, "Scientism and Technology as Religions," *Zygon*, 40:4: 835–844 (December 2005) 836.

[15] Paul Davies, *God and the New Physics* (New York: Simon and Schuster, 1983) 16.

[16] Frank Tipler, *The Physics of Immortality* (New York: Doubleday, 1994) ix, 10, 17, 247.

[17] E. O. Wilson, *On Human Nature* (New York: Bantam, 1978).

assumption about the human condition and in time uncover the bedrock of the moral and religious sentiments. The eventual result of the competition between the two world views, I believe, will be the secularization of the human epic and of religion itself."[18] Here religion is defeated in the war by conquering and colonizing it.

Scientific imperialism may appear to the preacher like the whore of Babylon. It appears at first to be attractive, because on the face of it scientists say good things about religion. But its cup of abominations is seductive. Once within its grip, what is precious to the Christian disappears: a transcendent God, an active God, a gracious God, and a God who is capable of redemption. Prophetic critique is called for.

The Third Warfare Model: Ecclesiastical Authoritarianism

Now we turn to warfare from the point of view of the other army. *Ecclesiastical authoritarianism*, our third model, is what every scientist fears from the church. According to this model, modern science clashes with religious dogma that is authoritatively supported by ecclesiastical fiat, the Bible, or in Islam by the *Qur'ān*. The best example is the 1864 *Syllabus of Errors*, promulgated by the Vatican. Here it is asserted that scientific claims must be subject to the authority of divine revelation as the church has discerned it. Ecclesiastical authoritarianism wins the war over science through intellectual intimidation.

The Second Vatican Council in 1962–65 reversed this, affirming academic freedom for natural science and other secular disciplines. "This sacred Synod affirms the legitimate autonomy of human culture and especially of the sciences."[19] At the present time many of the leaders in the dialogue between science and faith engage one another at the Vatican Observatory or in the Pontifical Academy of Sciences. Roman Catholic scientist Jerry D. Korsmeyer, while trying to develop a "theology of evolution," writes, "there is no opposition between science and theology."[20]

The 21st century pastor is not likely to be tempted to defend church dogma against an alleged scientific assault as did the Roman Catholic Church of the 19th century, to be sure. Yet, the integrity of theologically derived truths and observations needs to be explored without the anxiety that science will soon explain all such claims away. The parish leader, in this circumstance, needs to trust the truth – that is, trust that

---

[18] E.O. Wilson, *Consilience: The Unity of Knowledge* (New York: Alfred A. Knopf, 1998) 265.

[19] *Gaudium et Spes* [Pastoral Constitution on the Church in the Modern World] § 59 in *The Documents of Vatican II*, ed. by Walter M. Abbott, S.J. (New York: The America Press, 1966) 265.

[20] Jerry D. Korsmeyer, *Evolution and Eden* (New York: Paulist Press, 1998) 9.

open and fair discussion of any theological idea or any scientific idea will lead eventually to edifying truths and not to any embarrassment for either faith or science. Church teaching situations should embrace an atmosphere of confident openness and exploration.

The Fourth Warfare Model: The War Over Darwinian Evolution

The fourth of our warfare models is the *battle over Darwinian evolution*. The battlefields are churches, public school classrooms, school boards, university lecture halls, and the courts in North America, Australia, and Turkey, with little or no notice in Europe. Before picking sides and leaping into battle with guns blazing, the pastor should pause to see who is fighting with whom about what? So, we will pause here to provide more detail than we have for the other models.

Five fighting armies are discernable, making it much more complicated than the image of a simple war between science and religion might connote. The first position would be that of *evolutionary biology strictly as science* without any attached ideological commitments. The reigning theory is neo-Darwinian. Neo-Darwinism combines Charles Darwin's original nineteenth-century concept of natural selection with the twentieth-century concept of genetic mutation to explain the development of new species over 3.8 billion years. Defenders of quality science education in the public schools most frequently embrace this "science alone" approach.

The second position combines neo-Darwinism with the scientism mentioned above to formulate a *materialist ideology*. This ideology includes repudiation of any divine influence on the course of evolutionary development. Spokespersons for *sociobiology*, such as E.O. Wilson or Richard Dawkins, are aggressive and vociferous. Evolution here provides apparent scientific justification for scientism, scientific imperialism, and in some cases belligerent atheism. Charles Darwin himself did not draw atheistic implications from his science, writing, "My views are not at all necessarily atheistical."[21] But, his disciples do. "Darwin made it possible to be an intellectually fulfilled atheist," writes Oxford's Richard Dawkins.[22]

The third position is *Scientific Creationism*. During the fundamentalist era of the 1920s, biblical creationists appealed to the authority of the Bible to combat the rise in influence of Darwinism. Since the

---

[21] Charles Darwin, Charles, *The Life and Letters of Charles Darwin, Including an Autobiographical Chapter*, ed. by his son, Francis Darwin (3 Volumes: London: John Murray, 1888) 2:312.

[22] Richard Dawkins, *The Blind Watchmaker* (New York/London: W.W. Norton, 1987) 6.

1960s, creationists have based their arguments not on biblical authority but rather on counter science – hence their label, *Scientific Creationists.* They argue, for example, that the fossil record will contradict standard appeals to natural selection over long periods of time. Those known as Young Earth Creationists (YECs), such as the leaders of the Institute for Creation Research near San Diego, California, hold that the planet earth is less than ten thousand years old and that all species of plants and animals were originally created by God in their present form. They deny macroevolution – that is, they deny that one species has evolved from prior species; although they affirm microevolution – that is, evolution within a species. Key here is that creationists justify their arguments on scientific grounds. "Creation is true, evolution is false, and real science confirms this" writes Henry Morris.[23]

The fourth position is *Intelligent Design* or ID. Advocates of Intelligent Design sharply attack neo-Darwinian theory for overstating the role of natural selection in species formation. They argue that slow incremental changes due to mutations are insufficient to explain the emergence of new and more complex biological systems. Many of the life forms that have evolved are irreducibly complex, and this counts as evidence that they have been intelligently designed. Intelligent Design scholars such as Michael Behe, Philip Johnson, and William Dembski posit that appeal to a transcendent designer is necessary for the theory of evolution to successfully explain the development of life forms. Here scientific questions lead to theological answers.

The fifth position is *theistic evolution*, according to which God employs evolutionary processes over deep time to bring about the human race and perhaps even carry the natural world to a redemptive future. Theistic evolution first appeared in the late nineteenth and early twentieth centuries, even in the work of conservative Princeton theologian B.B. Warfield, for whom God's *concursus* with nature brought about the human race, just as God's *concursus* wrote the Scripture with human minds and hands. Teilhard de Chardin is perhaps best known for his evolutionary cosmology directed by God toward a future "Point Omega." Among contemporary scholars at work in the field of science and religion, the roster of theistic evolutionists includes Arthur Peacocke, Philip Hefner, Robert John Russell, Nancey Murphy, Kenneth Miller, John Haught, Martinez Hewlett, and Howard van Til. This school of thought is not occupied with defending evolution against attacks by advocates of scientific creationism or Intelligent Design;

[23] Henry Morris, *History of Modern Creatonism.* (Santee CA: Institute for Creation Research, ^{2}1993) 308–309.

rather, it seeks to work through questions raised by randomness and chance in natural selection in light of divine purposes and ends.[24]

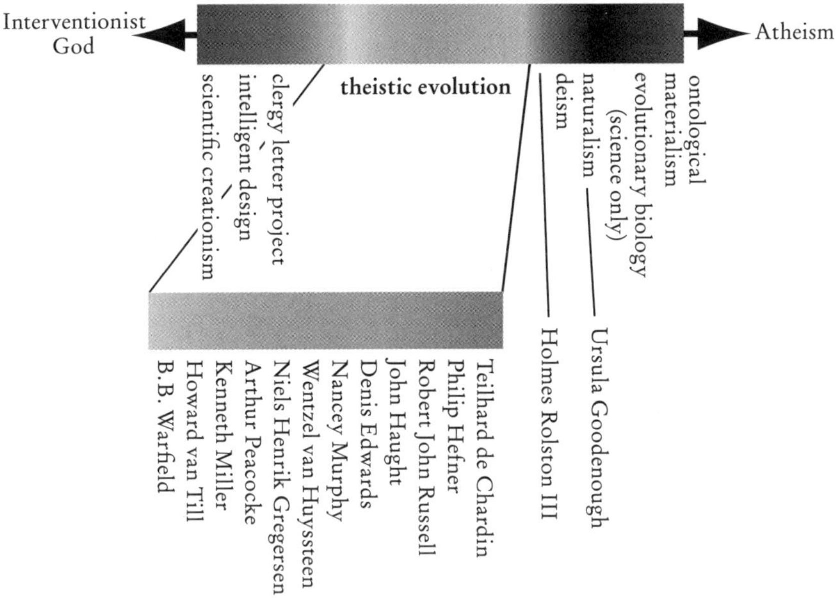

Figure 6: Theistic Evolution by Martinez Hewlett. Used with permission.

That a war is being fought is clear. However, because the actual points at issue deal specifically with the explanatory adequacy of natural selection, it would be misleading to simply dub this a war between science and religion. We here stress the point that what the parish pastor needs to know is this: despite the fact that superficially this looks like a battle between science and faith, down deep it is not. All combatants revere and respect science. It is in essence a battle over what constitutes good science. A parish leader who views this as a battle of science versus faith and who takes the side of faith does so only at grave peril.

What the pastor needs to stress in the congregational setting is that the Christian faith need not identify with anti-Darwinism, when Darwinism is understood as good healthy science and not materialistic

[24] See: Ted Peters/Martinez Hewlett, *Evolution from Creation to New Creation* (Nashville: Abingdon Press, 2003) and Robert John Russell/William R. Stoeger/ Francisco J. Ayala, *Evolutionary and Molecular Biology: Scientific Perspectives on Divine Action,* (Vatican City State and Berkeley: Vatican Observatory and CTNS, 1998).

ideology. The Christian faith opposes materialistic ideology, but it strongly supports good healthy science.

The impact of this war on young people should be a matter of pastoral concern. If our youth get the impression that a competition exists between faith and science, they may eschew the possibility of a career in science. When they get to the university they may avoid courses in biology. This would be as tragic as it is unnecessary. The pursuit of truth about nature through the study of science ought to be considered a Christian vocation. The parish pastor should from time to time lift up the vocation of science as something to emulate and celebrate, so our young people can consider it a possible future for themselves.

Without being overly formulaic, we might say that the four warfare models just described all deserve a critical deconstruction and a prophetic warning. Each is misleading to a person of faith. The preacher needs to guard against any of them gaining tacit endorsement in his or her world of the unsaid. Each may from time to time deserve focused attention and critique.

The First of Four Non-Warfare Models: The Two Languages

We have just reviewed four ways of understanding the relationship between science and faith in terms of warfare: scientism, scientific imperialism, ecclesiastical authoritarianism, and the battle over evolution. But, warfare is not the only lens through which to look at this relationship. In fact, the image of warfare could be misleading. "An image of perennial conflict between science and religion is inappropriate as a guiding principle", writes Oxford's John Hedley Brooke.[25] In the contemporary globe wide conversations between scientists and religious thinkers, at least four non-warfare or cooperative models have emerged. Any or all of these could become material for the parish pastor to use in constructing a meaningful worldview that includes modern science.

The first and most widely embraced model for relating science and religion is the *two languages* model. According to this model, science speaks one language, the language of facts, and religion speaks a different language, the language of values. The Two Language model – sometimes referred to as the "independence" model – is the prevailing view of both scientists and theologians in Western and Asian intellectual life. Science attends to objective knowledge about objects in the penultimate realm, whereas religion attends to subjective knowledge about transcendent dimensions of ultimate concern. Modern persons need

[25] John Hedley Brooke, *Science and Religion: Some Historical Perspectives* (Cambridge: Cambridge University Press, 1991) 33

both, according to Albert Einstein, who claimed the following: "Science without religion is lame and religion without science is blind."[26] Warfare is avoided by establishing a border and keeping science and faith in their respective territories.

This Two Language model should not be confused with the classic model of the *Two Books*, according to which the book of Scripture and the book of nature each provide an avenue of revelation for God.[27] The difference is that the Two Books model sees science as revealing truth about God, whereas the Two Language model sees science as revealing truth solely about the created world.

Already many preachers rely on the two language model when expositing biblical texts such as the Genesis creation accounts. After grasping for a few possible connections or crossovers between Genesis and Big Bang cosmology or Darwinian evolution, the preacher typically elects to say that science tells us what happened while the Bible tells us what it means. This is perfectly legitimate. It is safe, and satisfactory. It does not risk losing any credibility. It provides an accessible safety zone within which to present the biblical message unencumbered with otherwise difficult to answer questions.

The Second Non-Warfare Model: Dialogue

For many scholars in actual dialogue between scientists and theologians, however, the two languages model for keeping the two independent of one another is inadequate. It is theoretically inadequate, from the point of view of the theologian. The theologian asks: if it is true that the God of Israel is the creator of this world; and if it is true that natural scientists are gaining accurate knowledge of how this world works; then, sooner or later, we would expect to see some convergence or at least consonance between the two domains of knowing.

Jürgen Moltmann, for example, complains about the lack of interaction between science and theology. "Today the dilemma between theology and science is no longer that they present conflicting statements. It is rather the lack of conflict between statements which stand side by side without any relation to one another ... They are resting side by side in a vacant co-existence."[28]

This leads to the next non-warfare model, dialogue. Dialogue refers to actual conversation between scientists and theologians about matters

---

[26] Albert Einstein, *Ideas and Opinions* (New York: Crown, 1954, 1982) 49.

[27] See: Peter M.J. Hess, "God's Two Books: Special Revelation and Natural Science in the Christian West," in *Bridging Science and Religion*, ed. by Ted Peters/Gaymon Bennett (Minneapolis: Fortress Press, 2003) chapter seven.

[28] Jürgen Moltmann, *Science and Wisdom* (Minneapolis: Fortress Press, 2003) 2.

of shared interest. Within the concept of dialogue we find a very important assumption, namely, *hypothetical consonance*. Going beyond the Two Language view by assuming an overlap between the subject matter of science and the subject matter of faith, *consonance* directs inquiry toward areas of correspondence between what can be said scientifically about the natural world and what can be said theologically about God's creation. Even though consonance seems to arise in some areas, such as the apparent correspondence of Big Bang cosmology with the doctrine of creation out of nothing, consonance has not been fully confirmed in all relevant shared areas. Therefore, the adjective *hypothetical* applies to theology as well as science. The key hypothesis of this model is that there can be only one shared domain of truth regarding the created world, and science at its best and faith at its best both humble themselves before truth. It follows, then, that one should trust that consonance will eventually emerge. Hypothetical consonance provides the basis for what some call "dialogue between science and theology" and others the "creative mutual interaction of science and theology."

The commitment to *creative mutual interaction* between science and theology has been most forcefully made by Robert John Russell, founder and director of the Center for Theology and the Natural Sciences at the Graduate Theological Union in Berkeley. Russell would like dialogue to lead to engagement and finally to interactive influence. Theology should take on board insights of science that are consonant with the picture of reality drawn by one's knowledge of God. When dissonance between what science says and what theology says arises, it should be faced honestly. Then, says Russell with considerable courage, theology should pose to science insights about reality that derive specifically from its knowledge of God. These insights should be translated into testable hypotheses, then given to the scientist to initiate a research program. If theology can provide a hypothesis to be tested generated from within theology's own reservoir of knowledge, then a genuine creative mutual interaction will be taking place.[29]

Although the parish pastor may intellectually presume hypothetical consonance, it will be difficult routinely to engage in creative mutual interaction with the scientific community. Paths simply do not cross frequently enough.

Also, very little in the typical sermon is of a hypothetical character. Hypotheses are for scholars who explore new ideas; sermons occasionally though rarely do this. Perhaps what could inform a sermon would be a background of unsaid trust in the truth, trust in the truth wherev-

[29] Robert John Russell, "Bridging Theology and Science: The CTNS Logo," *Theology and Science*, 1:1 (April 2003) 1.

er it may be found. The sermon could carry the mood that truth, even if unearthed by a scientist, could be celebrated by a person of faith.

The Third Non-Warfare Model: Ethical Overlap

The third non-warfare model and number seven on our comprehensive list is *ethical overlap*. Building on the Two Language model, wherein mutual respect between scientists and religious leaders is affirmed, some express a strong desire for religious cooperation on public policy issues deriving from science and technology. The ecological crisis and human values questions deriving from advances in genetics both enlist creative cooperation.

Pope John XXIII told us in *Pacem in terris* that Roman Catholics could make partners with "all persons of good will" when working for world peace. Working with all persons of good will in the community or around the world – even scientists of good will – to make our planet a better place in every respect ought to be the daily diet of any Christian congregation.

Today's parish pastor regularly looks for secular partners and partners in other religious traditions with whom to cooperate ethically in making our world a better place. Whenever and wherever we find morally minded allies we should take advantage. What this implies is that in preaching and teaching the pastor should look for opportunities to lift up the commitments of selected members of the scientific community who are striving to make our world a better place.

The Fourth Non-Warfare Model: New Age Spirituality

This brings us to the final example on our list, number eight, *New Age spirituality*. Having left the conflict or warfare model behind, synthetic spiritualities, such as those found in the New Age movement, seek to construct a worldview that integrates and harmonizes science with religion. Evolution becomes an overarching concept that incorporates the sense of deep time and imbues the development of a global spiritual consciousness as an evolutionary advance for the cosmos. Many here are prompted by the visionary theology of Teilhard de Chardin, although this Jesuit forerunner could not himself be categorized as New Age. Others in the New Age movement seek to integrate the experience of mystery articulated in Hinduism and Buddhism with advanced discoveries in physics, such as indeterminacy and quantum theory.[30]

The New Age concept of "conscious evolution," mentioned in an earlier chapter, will provide an example of the grand vision accompanied by confidence in spiritual progress. "The ultimate purpose of *Conscious*

[30] See: Ted Peters, *The Cosmic Self* (San Francisco: Harper, 1991).

Evolution as a world view is to foster the evolution of our species to full potential, based on the harmonious use of all our powers – spiritual, social, and scientific – in harmony with the deeper patterns of nature and the Great Creating Process itself, traditionally called God."[31]

The metaphysics of New Age will attract some pastors and repel others. Although some Western scientists and many Eastern scientists resonate with New Age spirituality, it is generally held in academic disrepute because it lacks the rigor found in both scientific research and classical Christian theology. It would be a mistake for the preacher to ride the New Age horse, because it will lose in the intellectual credibility race. Ethical overlap with special emphasis on ecology would be the domain most likely shared by New Agers and Christian pastors.

Here is a summary chart on warfare and non-warfare models of the relationship between science and theology.

	Warfare	*Non-Warfare*
Scientism	✗	
Scientific Imperialism	✗	
Ecclesiastical Authoritarianism	✗	
Evolution vs. Creationism	✗	
Two Languages		✗
Dialogue		✗
Ethical Overlap		✗
New Age Spirituality		✗

Figure 7: Evolution Options. Ted Peters.

The Problem of Scientists Playing God

One dramatic component to the reigning scientific worldview is mythical in the classical sense, namely, the concept of "playing God."[32] In fact, the concept of playing God is itself a survival of the classical Greek myth of Prometheus. In the modern scientific setting, this myth takes the thinly disguised form of Faust, Frankenstein, or *Jurassic Park*. The parish preacher and teacher needs to be aware of the actual myths

[31] The Foundation for Conscious Evolution at: http://www.evolve.org/pub/doc/evolve_what_is_ce.html.

[32] See: Ted Peters, *Playing God? Genetic Determinism and Human Freedom* (New York/London: Routledge, rev. ed., 2002).

within our modern worldview, so that demythologizing the scientific myth can become part of the hermeneutical process.

Public policy controversies, as we all can see, invoke abhorrence to "playing God." The phrase "playing God" refers to the power that science confers upon the human race to understand and to control the natural world. Even though the phrase, "playing God," is by no means a theological phrase, it sounds like one; and people within the churches will ask about its meaning and significance. The pastor needs to be prepared to answer.

Upon close examination, the widely used phrase "playing God" seems to connote three overlapping meanings. The first meaning is connected to basic scientific research: to play God is to *learn God's awesome secrets*. Some scientific discoveries evoke a sense of awe and wonder over the complexity and majesty of the natural world that the human mind is apprehending. To pursue science is like shining a light into darkness. It makes visible the secrets of nature. It reveals what previously was hidden. It is the revelatory power of science that leads us to think we are gaining godlike knowledge. Few would ask us to cease our investigation, because "learning for learning's sake" remains the morality of scientific knowledge.

The second meaning of "playing God" is found in the hospital. It connotes that doctors have gained the *power over life and death*. When overwhelmed by a medical emergency, as a patient we feel helpless, totally dependent upon the scientific training and personal skill of the physician attending. The surgeon, and the scientific training he or she has been exposed to in medical school, stands between us and death. Similarly, large scale research programs dedicated to finding a cure for cancer or AIDS provide the larger society with hope in the face of helplessness. Here "playing God" takes on a redemptive or salvific significance. The genre of jokes about doctors who think of themselves as gods reflects the wider anxiety we have over helplessness, combined with our dependence upon doctors and their skills.

When we use "playing God" in this medical sense, we make two assumptions. First, we assume that decisions regarding life and death belong to God's prerogative. The second assumption follows from the first: when we find a human being with the power of life and death, we think of him or her in a godlike role. This provokes additional anxiety. We worry that the person in the godlike role will succumb to the temptation of pride, of hubris. The concept of hubris articulates the more inchoate fear that we will presume too much, overreach ourselves, violate some divinely appointed limit, and reap destruction. Anxiety over hubris marks the overlapping transition from the second to the third use of the phrase "playing God."

The third meaning of "playing God" refers to our ability to *alter life and influence human evolution*. This meaning is best expressed in the story of Frankenstein, the mad scientist who violates an invisible boundary and crosses over into the sacred realm of nature; then nature rises up vengefully and unleashes death dealing chaos. Our society fears the mythical mad scientist, who by violating nature may cause a backlash that will lead to suffering on the part of us all.

We assume here that scientific understanding leads to technological control. We want control. Yet, we doubt our own wisdom to know how to use this control. In our attempt to gain control over nature, we may so violate nature that it will lash back with destructive force. This fear is most associated with genetics and ecology. Genetic engineering, wherein we alter our genome and perhaps alter our own essence, is the primary area of science that provokes fears of playing God. Such fears also arise in ecology, where we worry that civilization may soon pass the point of no return and the environment will poison the human race into extinction.

We have just described the cultural meaning of the phrase, "playing God." This term belongs in the category of cultural myth. It provides a cipher for discerning the love and fear relationship our wider society has with the scientists among us. The parish pastor dare not misinterpret what is transpiring here. The pastor dare not simply respond to shrill voices of anxiety or hysteria that denounce godless science for endangering the human race with its Frankensteinian recklessness. The pastor needs to pause, analyze, discern, and reflect, before responding. But, a response is definitely called for.

The first thing to observe is that the God of "playing God" is not necessarily the God of the Bible. Rather, it is divinized nature. In Western culture nature has absorbed the qualities of sacredness. Science along with technology risk profaning the sacred.

We have suggested above that we are talking here about myth, about the place of science in the contemporary worldview. In this case, an actual mythical story is at work. Our use of the phrase "playing God" relies on the ancient Greek myth of Prometheus. According to this myth, when the world was being created, the sky god Zeus was in a cranky mood. The top god in the Olympiad decided to withhold fire from earth's inhabitants, leaving the nascent human race to relentless cold and darkness. Prometheus the Titan, whose name means "to think ahead," could foresee the value of fire for warming homes and providing lamplight for reading late at night. He could anticipate how fire could distinguish humanity from the beasts, making it possible to forge tools. So, Prometheus craftily snuck up into the heavens where the gods dwell and where the sun is kept. He lit his torch from the fires of the sun, then he carried this heavenly gift back to earth.

The gods on Mount Olympus were outraged that the stronghold of the immortals had been penetrated and robbed. Zeus was particularly angry over Prometheus's *hubris*, so he exacted merciless punishment on the rebel. Zeus chained Prometheus to a rock where an eagle could feast all day long on the Titan's liver. The head of the pantheon cursed the future-oriented Prometheus: "Forever shall the intolerable present grind you down." The moral of the story, which is remembered to the present day, is this: human pride or hubris that leads us to overestimate ourselves and enter the realm of the sacred will precipitate vengeful destruction. The Bible provides a version of the same point: "Pride goes before destruction" (Proverbs 16:18).

For us in the modern world who think scientifically, no longer does Zeus play the role of the sacred. Nature has replaced the Greek gods. It is nature who will strike back in the Frankenstein legend or its more contemporary geneticized version, Michael Crichton's novel *Jurassic Park* (1990) and the subsequent movies. The theme has become a common one: the mad scientist exploits a new discovery, crosses the line between life and death, and then nature strikes back with chaos and destruction.

Interpreting the Gene Myth

How should religious leaders interpret this classic myth as it influences contemporary culture? Should the parish pastor believe the myth? Should the parish pastor act out of the worldview of this myth? Or, should the pastor demythologize?

Numerous religious leaders have allowed themselves to conform to the myth; and they have taken up rhetorical arms against science. A 1980 task force report, *Human Life and the New Genetics*, includes a warning by the U.S. National Council of Churches: "Human beings have an ability to do Godlike things: to exercise creativity, to direct and redirect processes of nature. But the warnings also imply that these powers may be used rashly, that it may be better for people to remember that they are creatures and not gods." A United Methodist Church Genetic Science Task Force report to the 1992 General Conference stated similarly: "The image of God, in which humanity is created, confers both power and responsibility to use power as God does: neither by coercion nor tyranny, but by love. Failure to accept limits by rejecting or ignoring accountability to God and interdependency with the whole of creation is the essence of sin." In the hands of Christian leaders, the myth tells us that we can sin through science by failing to recognize our limits and, thereby, violate the sacred.

However, there is an alternative route one can take. As we noted above, genetics (along with ecology) is the field of research that provokes the most anxiety regarding the threat that scientists will play God. This is because DNA has garnered cultural reverence. The human genome has become tacitly identified with the essence of what is human. A person's individuality, identity, and dignity have become connected to his or her individual genome. Therefore, if we have the hubris to intervene in the human genome, we risk violating something sacred. This tacit belief is called by some the "gene myth," by others "the strong genetic principle" or "genetic essentialism." This mytheme – a mini-myth within the larger Promethean myth within the still larger worldview of science – provides an interpretive framework that includes both the assumed sacrality of the human genome combined with the fear of Promethean pride.

The preacher and teacher in the parish setting should critically if not prophetically question the gene myth. Doubt should be registered about the equation of DNA with human essence or human personhood. A person is more than his or her genetic code. The National Council of Churches of Singapore put it this way in *A Christian Response to the Life Sciences*: "It is a fallacy of genetic determinism to equate the genetic makeup of a person with the person."[33] No person is reducible to his or her genome. No person is a victim of a thoroughgoing genetic determinism. At some level, this cultural myth needs deymythologizing if not descientizing, if the parish pastor is to move people to a reasonable and healthy understanding of human nature in light of our faith in God.

Genetics, Ethics, and Worldview Construction

In the pastoral setting, increasingly parishioners will come to their clergy for counsel and advice on genetic issues. Initially, couples planning to bring children into the world will visit with their genetic counselor at the clinic and then show up on the pastor's doorstep for further discussion. Pastors will need to understand the overlap between pastoral and ethical concerns that will come in a single package. Stem cell therapy, selecting genes for future children, altering genomes, aborting defective fetuses, and envisioning a genetic future for children will appear on the list of concerns. The parish pastor needs to be ready.

---

[33] National Council of Churches of Singapore, 2002, *A Christian Response to the Life Sciences*, 81.

Here, we will return to a topic discussed earlier, an issue that might confront the pastor with both counseling and ethical concerns, namely, the relationship between genetic therapy and genetic enhancement. In addition to appearing as a pastoral matter, it is also a public policy issue.

Employment of advancing genetic technologies to alter human DNA leads to considerations regarding the distinction between therapy and enhancement. At first glance, therapy seems ethically justifiable, whereas enhancement seems Promethean and dangerous. The term *gene therapy* refers to directed genetic change of human somatic cells to treat a genetic disease or defect in a living person. With four to six thousand human diseases traceable to genetic predispositions – cystic fibrosis, Huntington's disease, Alzheimer's, many cancers – the prospects of gene based therapies are raising hopes for dramatic new medical advances. Few if anyone find ethical grounds to prohibit somatic cell therapy via gene manipulation.

The term *human genetic enhancement* refers to the use of genetic knowledge and technology to bring about improvements in the capacities of living persons, in embryos, or in future generations. Enhancement might be accomplished in one of two ways: either through genetic selection during screening or through directed genetic change. Genetic selection may take place at the gamete stage, or more commonly as embryo selection during preimplantation genetic diagnosis (PGD) following in vitro fertilization (IVF). Genetic changes could be introduced into early embryos, thereby influencing a living individual, or by altering the germ line, influencing future generations.

Some forms of enhancement are becoming possible. For example, introduction of the gene for IGF-1 into muscle cells results in increased muscle strength and health. Such a procedure would be valuable as a therapy, to be sure; yet, it lends itself to availability for enhancement as well. For those who daydream of so-called designer babies, the list of traits to be enhanced would likely include increased height or intelligence, as well as preferred eye or hair color. Concerns raised by both secular and religious ethicists focus on economic justice – that is, wealthy families are more likely to take advantage of genetic enhancement services, leading to a gap between the "genrich" and the "genpoor."

The most ethical heat to date has been generated over the possibility of germ line intervention, and this applies to both therapy and enhancement. The term *germ line intervention* refers to gene selection or gene change in the gametes, which in turn would influence the genomes of future generations. Because the mutant form of the gene that predisposes for cystic fibrosis has been located on chromosome four, we could imagine a plan to select out this gene and spare future generations of the suffering caused by this debilitating disease. This would constitute germ line alteration for therapeutic motives. Similarly, in principle, we could select or even engineer genetic predispositions to favorable traits in the same manor. This would constitute germ line alteration for enhancement motives.

Such efforts at genetic engineering are risky. Too much remains unknown about gene function. It is more than likely that gene expression works in delicate systems, so that it is rare that a single gene is responsible for a single phenotypical expression. If we remove or engineer one or two genes, we may unknowingly upset an entire system of gene interaction that could lead to unfortunate consequences. The prohibition against "playing God" serves here as a warning to avoid rushing in prematurely with what appears to be an improvement but could turn out to be a disaster. Ethicists frequently appeal to the precautionary principle – that is, to refrain from germ line modification until the scope of our knowledge is adequate to cover all possible contingencies.

It is important to note that the precautionary principle does not rely upon the tacit belief in DNA as sacred. Rather, it relies upon a principle of prudence that respects the complexity of the natural world and the finite limits of human knowledge.

Conclusion

The parish pastor or priest could be motivated to invest time and energy in the worldwide dialogue between science and theology out of pure curiosity, or even out of a desire to enhance the effectiveness of his or her own ministry. We have tried to show here that three areas of pastoral responsibility – preaching, teaching, and counseling – all could benefit from increased sophistication regarding the scientific picture of the world in which we live. By no means do we suggest that science provide the content of what the preacher or teacher or counselor has to say; yet, for the sake of credibility and relevance, what is said should resonate with what is unsaid about the scientific worldview we all share.

Chapter 9

Proleptic Dignity, Proleptic Ecology, and Prophetic Politics

"Where do you draw the lines?" is a question that is asked of ethicists again and again by journalists, religious leaders, and even by Roman Catholic and Protestant ethicists themselves. I simply do not like this question. Much less do I like ethicists to answer it.

Hidden in the question is a disparaging assumption about the task of ethics. Ethics, it is assumed, has the job of drawing lines, of erecting fences, of placing barriers and "no trespassing" signs to prevent people from moving forward. The only two words an ethicist apparently needs to utter are "no" and "stop." This is strictly a negative pre-understanding of the purpose and task of ethical deliberation.

Can an ethicist help anybody? Can an ethicist help anybody get from one situation to a better situation? Can an ethicist help anybody who feels the need for healing, for redemption, for transformation, for fulfillment?

Ethicists are concerned about what is good. The good is that for which we strive but do not yet possess. A vision of what is good draws us from the present situation toward something that is better, from a situation of lack toward one of fulfillment. To get from here to there may require transformation. Could an ethicist help us envision transformation?

"The good ... [is] that at which all things aim,"[1] said Aristotle. Two implications are worth noting here. First, the good is not what we have; rather, it is what we aim for. The good has a built in future orientation. Second, as the aim or goal or *telos*, the good defines itself as an end and our actions as the means. The end defines the means. "We may define a good thing as that which ought to be chosen for its own sake."[2] In turn, our moral actions become defined as good insofar as they become means for reaching the end.

The task of the ethicist, it seems to me, is to lift up and make visible the aim, and to encourage pursuit of transformation in light of the aim.

---

[1] Aristotle, *Nicomachean Ethics*, I: 1.

[2] Aristotle, *Rhetoric*, I: 6.

Rather than say "no," the ethicist should say "go." Rather than say "stop," the ethicist should say "hop."

In what follows I would like to advance three theses, implications of a proleptic approach to ethics. The first is the *Proleptic Dignity* Thesis: Human dignity belongs at the center of our value system; and we should understand dignity as first conferred, then inherent. The second is the *Proleptic Ecology* Thesis: In response to the urgency of the eco-crisis, we should orient ethical action around our eschatological vision of a justly ordered creation at peace; and we should take action to transform present destructive trends into constructive attempts at transformation. The third is the *Prophetic Proviso* Thesis: The eschatological vision needs an accompanying prophetic proviso – that is, inspiration for transformation accompanied by a precautionary warning against immanentizing the transcendent.

Far more than limiting ethics to saying "no" or "stop," proleptic ethics just might be of help to people striving to make this a better world.

1. The Proleptic Dignity *Thesis:* *Human dignity belongs at the center of our value system; and we should understand dignity as first conferred, then inherent.*

A human person is a good in Aristotle's sense: an end to be chosen for its own sake. A human person is not a means to some additional end, but an end in himself or herself. This application of the concept of the good to an individual person we refer to as *dignity*. In contemporary ethics, human dignity is like a Visa credit card; we would not go anywhere without it.

However, we need to sort through a conceptual problem, an ontological problem. Since the Enlightenment, we have begun to think of dignity as an inherent attribute, a state of being with which we are born. Dignity appears to be intrinsic, and the recognition of a person's dignity appears to be the recognition of an innate quality. What has happened culturally is that we have reified or hypostacized the reason for treating a person as an end, so that dignity has become an inherent attribute of the person. Treating a person as an end has led to a presumed ontology of individual dignity. What is missing are two conceptual factors necessary for fully understanding dignity, namely, that dignity is future and dignity is relational.

Why is dignity future? Because dignity is a good we seek, not one we already possess. It lies beyond the present state of our being. We might place dignity within the larger quest for the good when admitting what Wolfhart Pannenberg asserts. "The good is that which man does not yet possess conclusively, that which he must still strive to realize. The good thus contains in itself the difference between value and being. The good is beyond being, Plato said."[3] Dignity does not have being today; but it will tomorrow. To treat persons with dignity today is grounded in our eschatological anticipation.

Why is dignity relational? Because it is the product of an interaction of the relationship of an individual with others, and with God. The dynamic that produces dignity begins with someone treating a person as a person – that is, treating a person as valuable, as worthy, as an end and not merely as a means. Once treated as a person of worth or final value, the person is enabled to rise up and claim that sense of worth or value for his or her own. Dignity is a product of love; and love is a quality of relationship. Let us look first at relationality; then we will turn to the proleptic character of dignity.

Dignity as Individual, Universal, and Relational

The cultures of the world have not always applied the concept of dignity universally, to all persons; and many if not the majority of persons in history have not been treated with dignity. In ancient Rome, only the elite classes enjoyed social or political dignity. *Dignitas* was ascribed to a wealthy member of the royal family or someone in a position of rank in society. The elite would be given respect and honor in public. Peasants and servants would have to defer, give place, run errands, pay obeisance, and perform whatever tasks are necessary to make the life of the person with *dignitas* secure and comfortable. Dignity distinguished the powerful in society from the marginal, those who get served from those who serve them.

This elitism was rejected by the wannabe Greeks of the 18th century European Enlightenment. We are heirs of Enlightenment anti-elitist egalitarianism. Now, dignity is in principle applied to all persons regardless of social status. The flowering of our modern Enlightenment notion of dignity is rooted in Greek philosophy and the Bible. The Greeks tilled the conceptual soil for the later growth of egalitarianism. They saw each human being as having an inherent capacity to reason, and reason opens the individual mind to the divine mind. This turn inward toward the mind in order to establish a connection with God eventually fertilized the growth of dignity for all persons.

Immanuel Kant speaks for the Enlightenment age. "Rational beings ... are called *persons* because their nature already marks them out as ends in themselves – that is, as something which ought not to be used

---

[3] Wolfhart Pannenberg, *Theology and the Kingdom of God* (Louisville: Westminter John Knox Press, 1969) 106.

merely as a means – and consequently imposes to that extent a limit on all arbitrary treatment of them ... Persons, therefore, are not merely subjective ends whose existence as an object of our actions has a value *for us*; they are *objective ends* – that is, things whose existence is in itself an end ... for unless this is so, nothing at all of *absolute* value could be found anywhere."[4] The modern (not postmodern) notion of universal human dignity is based upon the ancient Greek assumption of an inward capacity, the capacity of each person to reason. It is belief in the inborn capacity for rationality that, once let loose in Western culture, eventually led to the application of dignity not merely to the elite but to every reasonable person.

However, before dignity could mature into a universal, the turn inward needed an outward complement, namely, love. This brings us to the Christian contribution. But, you might ask: is not our concept of dignity rooted in the fact that human beings are made in God's image? Why love?

Yes, it is easy to understand why one might think that what we know as dignity is rooted in the biblical notion of the image of God, the *imago dei.* Adam and Eve represent universal humanity; so their creation in God's image implies that we all enjoy this divine image. To trace dignity to the origins of the human race in the *imago dei* (Genesis 1:26–29; 2:7), however, risks rendering dignity dependent on an innate quality that makes the human person godlike. The Adamic *imago dei* looks a bit like the Greek reliance on our inner capacity to reason.

Therefore, I do not believe the *imago dei* in this Adamic form is what has been decisive. Rather, what has been decisive is the unpredictable reversal in the relationship of Creator to creature; it is the role assumed by God that we creatures should be treated lovingly. God serves us as a peasant serves a king. "When the Bible calls human beings the image of God, this constitutes a fundamental criticism of the divinization of the rulers and their ideologies of rule. Not the king, but the individual human being alone is mediator between God and the people. Human beings do not exist for the sake of rule; rule, rather, exists for the sake of human beings," writes Jürgen Moltmann.[5]

In the German language, the word for Sunday morning worship is *Gottesdienst*, meaning "divine service." Martin Luther made a pun out of this. Instead of our serving God, he said, in worship we hear the word proclaimed that God has served us. The result is this: we who

---

[4] Immanuel Kant, *Groundwork of the Metaphysic of Morals*, tr. by H.J. Paton (New York: Harper, 1956) 96.

[5] Jürgen Moltmann, *On Human Dignity,* tr. by M. Douglas Meeks (Minneapolis: Fortress Press, 1984) 23.

were once humble are now exalted. If we are loved and respected by God, then we have dignity indeed.

The individual relation to God needed a communal interaction characterized by love, a relation in which each person regardless of social status is loved as an end and not as a means. Reason coming from within and love coming from without together fertilized and watered leading to an eventual harvest of dignity for all persons. The Greeks gave us reason. The Christians gave us love.

With Christianity, each person is beloved regardless of status. In fact, to treat a person lovingly is to ascribe to him or her such worth or value that our own life could be sacrificed on their behalf. The sacrifice of Jesus' life, Matthew 20:28 "the Son of Man came not to be served but to serve, and to give his life a ransom for many", established that even God values persons to this extent.

Adolph von Harnack described the Christian innovation in terms of the infinite value of the human soul. "The idea of the inestimable inherent value of every individual human soul, already dimly appearing in several psalms, and discerned by Greek Philosophers, though as a rule developed in contradiction to religion, stands out plainly in the preaching of Jesus. It is united with the idea of God as Father, and is the complement to the message of the communion of brethren realizing itself in love. In this sense the Gospel is at once profoundly individualistic and Socialistic."[6] Dignity is the result of a relationship of love; and the Christian community is charged by God to love in such a way that persons everywhere experience being treated as a beloved end, as a good in themselves.

When Christianity appeared on the scene, the dignity that previously belonged to only the select elite was given to each member of the human race, regardless of their capacity to reason. God became the servant in Jesus Christ, paying obeisance to those whom God loves. Jesus came not to be served, but to serve, he said. Thereby, God's incarnation dignifies us, all of us and each of us, making us an end to which divine humiliation in Jesus Christ is the means.

Like a germinating seed, the idea of universal dignity sprouted and flowered with implications such as human equality and liberation for those living under undignified circumstances. Love for those on the margin and compassion for those who are suffering became hallmarks of Christian ethics. These values gave birth in the Enlightenment to a new generation of ideals such as equality, democracy, justice, and human rights. Dignity today is a secular ethical principle with roots

---

[6] Adolph Harnack, *History of Dogma*, (7 Volumes: New York: Dover, [3]1900) 1:70.

in the Christian understanding of the human person as an everlasting object of God's love.

In our modern (not postmodern) post-Enlightenment period, we have assumed that the concept of dignity refers us to the intrinsic value of every human person. The value of a person cannot be reduced to his or her instrumental worth. This implies that we are always worth more than our possessions or our reputations or our function in the economy. As persons with dignity no one dare reduce us to the subjective value of those who like or dislike us. We are confident we can claim our rights even when everyone around dislikes us. Dignity applies to individual persons. And as individual persons we are always an end and never merely a means to some other greater value. It is this dimension of inherent value that constitutes human dignity as we know it in the modern West.

The value and significance of dignity understood as intrinsic or inherent or innate for our legal system is obvious. Such a dignity doctrine permits us in court to defend the rights of every individual regardless of how humble he or she might be. Every person has a right to *dignitas* before the law. We don't want to surrender this.

Still, we might ask, just what makes dignity inherent? Why do we assume it to be intrinsic, inherent, or innate? To get at this, let us ask a phenomenological question: so we actually experience dignity as inherent, or is it the result of being conferred? It is both. When young children grow up in a family where they receive love, they grow in self-worth. When their parents treat them as valuable, they begin to see themselves as valuable. Eventually, they claim that value for themselves. Phenomenally, dignity is first conferred, then claimed.

It is observable that love as an activity produces consequences. Love evokes a sense of dignity in the humble. To be the object of someone's love is to be made to feel valuable, to gain a sense of self-worth. Once we feel this sense of worth imputed to us by the one who loves us, we begin to own it. We begin to claim it. Worth is first imputed, then it is claimed.

From a theological perspective, our human dignity is ultimately conferred by God. Because in faith we experience God treating each of us with dignity, we now confer it on one another. Once we have conferred dignity on someone we love, we treat that person as having intrinsic value. The religious pursuit of justice by seeking the liberation of the oppressed from the cycle of poverty provides a dramatic form of conferring dignity. It is generated by love. This is the character of love, namely, to treat the beloved as an end and not a means to some further end.

Dignity is the result of grace, divine grace. For us to enjoy dignity, we must be loved by someone who is other. For us to confer dignity, we must love someone who is other. Elizabeth Johnson says the human self is "rightly structured not in dualistic opposition to the other but in intrinsic relationship with the other."[7] Dignity is relational before it becomes inherent. Dignity is first conferred relationally, then it is claimed independently. In this sense, dignity is the result of grace. Richard M. Gula makes this clear: "As long as God offers divine love (i. e. grace), humans will ever remain God's image and enjoy a sacred dignity whether in sin or not, whether acting humanly or not." This dignity is the gift to us from God's love. It is reinforced by saying we can do nothing to warrant it or merit it. "Human dignity does not depend ultimately on human achievements, but on divine love."[8]

Human and Trinitarian Personhood

We are who we are as persons-in-relation, in relation to one another, to the creation, and to God. Theological anthropology is the correlate of trinitarian theology.[9] The Holy Trinity is constituted by persons-in-communion. Personhood results from the dialectic of individuation and participation, according to Paul Tillich. "When individualization reaches the perfect form which we call a 'person', participation reaches the perfect form which we call 'communion'."[10]

Christian anthropology and the doctrine of the Trinity mutually imply one another here. The concept of the person in humanity connected to the concept of person in the Trinity. "The trinitarian doctrine implies a communitarian understanding of being human," writes Gula. "The trinitarian vision sees that no one exists by oneself, but only in relationship to others. To be is to be in relationship."[11] The tie between trinitarian theology and anthropology leads the late Notre Dame theologian Catherine Mowry LaCugna to assert: "Person indicates relationship." She continues: "Trinitarian doctrine is therefore inescapably a theology of personhood."[12] Personhood both within the divine life and as it relates to creation is the condition for the possibility

---

[7] Elizabeth A. Johnson, *She Who Is: They Mystery of God in Feminist Theological Discourse* (New York: Crossroad, 1993) 68.

[8] Richard M. Gula, S. S., *Reason Informed by Faith* (New York: Paulist Press, 1989) 64–65.

[9] See: Ted Peters, *God as Trinity: Relationality and Temporality in Divine Life* (Louisville: Westminster John Knox Press, 1993) 34–37; 94–95.

[10] Paul Tillich, *Systematic Theology* (3 Volumes: Chicago: University of Chicago Press, 1951–1963) I: 176.

[11] Gula, *Reason Informed by Faith*, 65.

[12] Catherine Mowry LaCugna, "The Trinitarian Mystery of God," *Systematic Theology: Roman Catholic Perspectives*, ed. by Francis Schussler Fiorenza/John P. Galvin (2 Volumes: Minneapolis: Fortress Press, 1991) 1: 180.

of what will become human personhood. God has a double relational dimension to being, says Eberhard Jüngel. "This means that God can enter into relationship (*ad extra*) with another being (and just in this relationship his being can exist ontically *without* thereby being ontologically dependent on this other being), because God's being (*ad intra*) is a being *related to itself.* The doctrine of the Trinity is an attempt to think out the self-relatedness of God's being."[13] The self-relatedness of God makes possible the self-relatedness of human beings; the other-relatedness of God makes possible the other-relatedness of human beings. Divine personhood makes possible human personhood.[14]

Theologians such as Roman Catholic Karl Rahner and Lutheran Wolfhart Pannenberg develop anthropologies that incorporate ontological openness, openness to otherness and openness to the transcendent.[15] Similarly, Eastern Orthodox theologian John Zizioulas proposes that personhood requires an "openness of being," an ecstatic relationship beyond the individual. A person is a self in the process of transcending the boundaries of the self. This self-transcendence is the root of freedom. Only relationship makes such self-transcendence, and hence freedom, possible. On the one hand, a person is an integrated unity, a self. On the other hand, a self who fulfills personhood is ecstatically open for communion.[16]

What Zizioulas assumes here is that relationality is ontological. Applying first and foremost to the Holy Trinity, it also applies to life that transcends death. "There is no true being without communion," he writes; "Nothing exists as an individual."[17] When it comes to human

---

[13] Eberhard Jüngel, *The Doctrine of the Trinity* (Grand Rapids MI: Wm. B. Eerdmans, 1976) 99.

[14] This openness toward the other in relationship, especially toward God, is essential to Christian anthropology. This openness is key to understanding what we mean by the human soul. "Human beings are bodily creatures who have a fundamentally unlimited transcendentality and unlimited openness to being as such in knowledge and freedom." Karl Rahner, *Theological Investigations* (21 Volumes: London: Darton, Longman, and Todd, 1961–1976; New York: Seabury 1974–1976; New York: Crossroad, 1976–1988) XXI:42. "A being is the more itself the more it is open, the more it is in realtionship." Johann Auer/Joseph Ratzinger, *Dogmatic Theology 9: Eschatology* (Washington DC: Catholic University Press of America, 1988) 155.

[15] Karl Rahner describes the human person with "openness to everything and anything" in *Foundations of Christian Faith* (New York: Crossroad, 1978) 31–32. Wolfhart Pannenberg describes this as openness-to-the-world or *Weltoffenheit* in *Anthropology in Theological Perspective*, tr. by Matthew J. O'Connell (Louisville: Westminster John Knox Press, 1985) 35–42.

[16] John D. Zizioulas, "Human Capacity and Human Incapacity: A Theological Exploration of Personhood," *Scottish Journal of Theology*, 28:5 (October 1975) 408.

[17] John D. Zizioulas, *Being as Communion* (Crestwood NY: St. Vladimir's Seminary Press, 1993) 18.

persons, he distinguishes between our biological hypostasis and our ecclesial hypostasis. Biologically, we are born to die. We are physically determined to pursue individualization, which tragically results in isolation from the source of life. We are born as a "*hypostatic* fact, as a body, but this fact is interwoven with individuality and with death."[18] True personhood arises through a trans-biological communion with God that transforms our relationship to the physical world. Key here is the necessity for personhood to be in communion and also that this relationship transcends our biological origin.

In summary, dignity is the product of relationship, our relationship to a loving God mediated by our relationship to others who treat us as persons of worth and value. Our ethic is one of conferring dignity on those who do not daily experience dignity. This relational understanding of dignity redounds to the doctrine of the *imago dei*, so that the *imago dei* must be understood relationally, not just individually. Jürgen Moltmann is a theologian who sees the *imago dei* as relational. "*The Image of God Is Human Beings in All Their Relationships in Life*."[19]

Proleptic Personhood

We have just elaborated one of the two typically overlooked dimensions of dignity, namely, its relationality or communal dependence. Now we turn to its proleptic character. Conferring dignity today anticipates inherent dignity tomorrow. Inherent dignity is eschatological; whereas conferred dignity is proleptic. What do I mean by this?

Dignity is proleptic, that is, it is fundamentally future oriented. Zizioulas affirms this eschatological dimension: "*The truth and the ontology of the person belong to the future, are images of the future.*"[20] Complementary to this, Carl E. Braaten incorporates eschatological resurrection into the definition of what makes us human: "In raising Jesus from the dead, God incorporated the dimension of eschatological fulfillment into the definition of human being."[21] Persons whom

---

[18] Ibid., 52.

[19] Moltmann, *On Human Dignity*, 23, Moltmann's italics.

[20] Ibid., 62, Zizioulas' italics.

[21] Carl E. Braaten, "The Person of Jesus Christ," in *Christian Dogmatics*, ed. by Carl E. Braaten/Robert W. Jenson (2 Volumes: Minneapolis: Fortress Press, 1984) II:524. "In this destiny of Jesus is anticipated the destiny of humanity and of the world," writes Zachary Hayes, O.F.M., *Visions of the Future: A Study of Christian Eschatology* (Wilmington DE: Michael Glazier, 1989) 55. Ibid., 62, Zizioulas' *Christian Dogmatics*, ed. by Carl E. Braaten/Robert W. Jenson (2 Volumes: Minneapolis: Fortress Press, 1984) 1:524.

we know and love today are on the way, so to speak; they anticipate their full essence as human beings by anticipating their resurrection and unity with Christ within the divine life. Our present conferral of dignity is itself part of this anticipation, a prolepsis of our eternal value conferred upon us by the eternal God. Dignity is not originally innate; it is eschatologically and retroactively innate.

Now, here is the central point I would like to make: dignity derives more from destiny than from origin, more from our future than from our past. This is because our dignity is tied to the eschatological image of God (*imago dei,* in Latin). The true image of God befitting the human race is not found in the first Adam but in the New Adam, not in Genesis but in Revelation. Adam prefigures the essential humanity that is yet to come. Who we are essentially as participants in the *imago dei* includes resurrection. Jesus Christ is the authentic image of God (*eikon tou theou,* in Greek), according to the New Testament; and this includes the Easter resurrection. For us to participate in Christ is to participate in resurrection. Essence derives from eschatology.

We find the authentic image of God in Christ first, only secondarily in Adam and Eve. Karl Barth makes this point vividly when interpreting Romans 5, the passage comparing and contrasting Adam with Christ. Our "essential and original nature" is to be found "not in Adam but in Christ. In Adam we can only find it prefigured. Adam can therefore be interpreted only in the light of Christ and not the other way around."[22] Barth goes on to observe that death is only an intruder into life, an alien force within God's intended creation. Death, though unavoidable for Adam, does not belong to the essence of humanity. Life, eternal life, is God's intent.

James Childs places the *imago dei* in its eschatological context. "The image of God is an eschatological concept which refers to man's destiny for the fulfillment of perfect communion and union with God." Childs adds, "the becoming of man in the promise of this destiny is constitutive of his being."[23] We today confer dignity on persons based upon our understanding of God's promise of their future fulfillment. This makes dignity both eschatological and proleptic. "A prolepsis is the presence of the future in the present."[24]

This is where ethics gains ontological grounding. Our ethical actions are grounded in our vision of omega. The conferring of dignity on someone who does not yet in fact experience or claim it is a gesture

---

[22] Karl Barth, *Christ and Adam* (New York: Collier, 1952, 1956) 39–40.

[23] James M. Childs, Jr., *Christian Anthropology and Ethics* (Minneapolis: Fortress Press, 1978) 110.

[24] Ibid., 7.

of hope, an act that anticipates what we hope will be a future actuality. Our final dignity, from the point of view of the Christian faith, is eschatological; it accompanies our fulfillment of the image of God. Rather than something imparted at conception or at birth, dignity is the future end product of God's saving activity which we anticipate socially when we confer dignity on those who do not yet claim it. The ethics of God's kingdom in our time and in our place consists of conferring dignity and inviting persons to claim dignity as a prolepsis of its future fulfillment.

Jesus' commandment to love one another means, among other things, the treating of all persons as a good, as an end and not merely as a means. This confers or imputes dignity to each person in such a way that he or she may rise up and claim self-worth for himself or herself and share in the benefits of living together on this planet. Our contemporary ethical mandate – what many call "justice" – is this: we should confer dignity on human persons so that they may rise up to claim dignity for themselves.

2. The Proleptic Ecology *Thesis:*
In response to the urgency of the eco-crisis, we should orient ethical action around our eschatological vision of a justly ordered creation at peace; and we should take action to transform present destructive trends into constructive attempts at transformation.

A cynic might complain that the protection or promotion of human dignity is not the most urgent of ethical concerns at the present time. What is of utmost urgency, one could say, is the environmental crisis along with its concomitant disparities in the distribution of earth's resources. "We need to recognize that Earth's resources are finite. To waste our limited resources is to lose them in the foreseeable future, with potentially dire consequences for all regions and the world," writes Mikhail Gorbachev. "We must wake up to the dangers of climate change and devote more resources to the crucial search for energy alternatives."[25]

In light of this global challenge, why are we talking here about dignity? Large scale even global concerns seem to override concerns for the dignity and rights and interests of individuals. The problem with placing dignity at the ethical center is that it fosters individualism and, one might complain, individualism is the cause of our ecological problem. Rather than individualism, we need to embrace a communal ethic; and then leap from communitarianism to planetarianism.

---

[25] Mikhail S. Gobachev, Forward to *State of the World 2005*, ed. by Linda Starke for the Worldwatch Institute (New York: W. W. Norton, 2005) xviii.

Victor Ferkiss provides us with an example of a futurist who has made this claim. "What has to be done to prevent the destruction of humanity by war, universal poverty, or cultural and biological perversion?" asks Ferkiss. "The only hope lies in a revolution in political philosophy." That revolution in political philosophy would replace individualism with communalism. Ferkiss uses the term 'liberalism' to describe the culprit, namely, the Enlightenment emphasis on human dignity in its individual form. "The essence of humanity's current crisis is that we have allowed our collective destiny to be determined by the political philosophy usually called liberalism, which holds that the prime purpose of human society is to encourage individual self-aggrandizement. It engages human society in a fatal attempt to conquer rather than cooperate with nature, licensing technology as the unfettered instrument of that conquest."[26] Then comes the moral challenge to choose between utopia or oblivion: "The future is completely open to human decisionmaking, whether for good or ill."[27]

The problem is not individualism as implied by the modern notion of human dignity, in my judgment. Rather, the problem is that the political philosophies we need to replace lack a global scope for transformatory ethics that organizes society around a transcendent vision, a transcendent vision that stands in prophetic judgment against absolute claims for partial accomplishments. The desired new political philosophy requires making human dignity central, otherwise it will devolve into tyranny and wreak even more destruction.

Kant is right, I believe. Unless we affirm the human person as the teleological end of our ethics, no absolute value will remain to orient our culture. The imperative of ecological health must serve the health of individual persons; just as we must enlist the fervent enthusiasm of individual persons to work toward environmental sustainability.

Since the establishment of the World Future Society in 1967 and the first "Earth Day" in 1969, futurist prophets have been forecasting changes in the ecosphere that could jeopardize the health of human life and animal life and plant life.[28] Modern industrial and post-industrial societies whirr like giant polluting machines, spewing environmental degradation day and night, decade after decade. Like an ocean hurricane picking up speed, unbridled global growth is spinning us toward one or another collision. The confluence of at least five torna-

---

[26] Victor Ferkiss, *The Future of Technological Civilization* (New York: George Braziller, 1974) 6.

[27] Ibid., 7.

[28] The most reliable source of relevant data regarding present trends, in my judgment, is the Worldwatch Institute: http://www.worldwatch.org.

dos could produce a single devastating planetary storm of destruction: (1) unstoppable growth in human population; (2) relentless depletion of non-renewable natural resources; (3) ever expanding agricultural production; (4) ever expanding industrial production; and (5) constant increase in pollution production. This constitutes the world *problematiqué*. Sooner or later, the Club of Rome[29] warned us more than three decades ago, these forces will collide and a massive human die back will be the result. An ecocatastrophe will take no prisoners. It will not respect human dignity.

The futurists of the late 1960s through the 1970s, challenged the world like the biblical prophets challenged ancient Israel to make the right decisions. If we do nothing, current trends will insure the destruction of the life-giving potential of the planet. If we embrace a utopian vision of global peace and justice combined with sustainability, we will not only survive but leap forward in our evolutionary history. It is our choice, said R. Buckminster Fuller, a choice between "utopia or oblivion."[30]

Futurist thinking as it occurs in technological societies follows a predictable pattern. I describe this pattern as *understanding-decision-control* (u-d-c). Futurology enlists the analytical tools of scientists to understand past and current trends, which provide what is necessary to project future trends and draw up alternative scenarios. Once alternative scenarios have been drawn up, they can be compared. They can be evaluated according to desirability. At this point values enter the picture, and the more valuable scenarios are selected. Decisions get made regarding actions to be taken. The assumption is that once the right decisions are made, the resulting actions will give us some degree of control over our future.

What the futurist fears most is mass denial. Large scale cultural denial can be prompted by garden variety ignorance or, worse, by local vested interests in short term profits at the expense of long term planetary health and wellbeing. The eco-prophets of the 1970s were drowned out by the factory noises of the 1980s, by the cheers of the stock market profiteers of the 1990s, and by the bombs dropping after the turn to the new millennium. Armed with the u-d-c formula, futurists and ecologists believe we need to persuade the citizens of our world to see the prophesied clouds on the horizon and prepare by making long range decisions. Like the prophets of ancient Israel, today's futurists are frustrated at ears which will not listen.

---

[29] Donella H. Meadows et. al., *The Limits to Growth* (New York: Universe Books, 1972).

[30] R. Buckminster Fuller, *Utopia or Oblivion: the Prospects for Humanity* (New York: Overlook Press, 1969).

Despite this frustration, the ethicist like the futurist must press the prophetic case into the consciousness of our global community. The Christian futurist can rely not simply on picturing a negative future, a doomsday, which we must take action to avoid. The Christian futurist can also project a glorious future, a God sponsored future that will transcend history's debacles. To see our actions on behalf of a healthy planet now as a partial anticipation of God's promised eschatological future grounds aggressive and creative ethics in the divine plan.

In summary, I recommend that the Christian ethicist join with secular futurists and ecologists in common cause. The Christian ethicist need not embrace a doctrine of inevitable technological progress, nor embrace philosophical naturalism, nor just look for ways to say "no" or "stop," even if others do. Rather, the Christian ethicist will encourage us to press on toward a precautionary yet sensible advance toward a more fulfilling future within history that is inspired by God's promise for a reality that transcends history. What Christian ethics brings to the fraternity of ecologists and futurists is the confident vision that proleptic ethics can rely upon.

Prolepetic Ecology

What is called for in the face of the world *problematiqué* is for Christians to provide (1 Peter 3:15) "an accounting for the hope that is in" us. This is a challenge, because the world *problematiqué* suggests hopelessness, at least according to John Polkinghorne and Michael Welker. "In the face of the environmental crisis, the continuing surge of global poverty, and the threat by an age of increasing conflict, scarcity, and despair, many people around the world look to a future without hope or joy."[31] Proleptic ethics can provide the accounting for hope and perhaps also provide a positive vision in the face of crisis.

Proeleptic ethics begins with projecting a vision of the historical future based upon biblical promises for an eschatological future. One such eschatological promise appears in Isaiah and has been nicknamed the "Peaceable Kingdom." It is a naturalistic vision of Edenic harmony, of a holistic ecology beyond the predator-prey violence of our evolutionary history.

> Isaiah 11:2–9. The spirit of the LORD shall rest on him, the spirit of wisdom and understanding, the spirit of counsel and might, the spirit of knowledge

---

[31] John Polkinghorne/Michael Welker, "Introduction," *The End of the World and the Ends of God: Science and Theology on Eschatology*, ed. by John Polkinghorne/Michael Welker (Harrisburg PA: Trinity Press International, 2000) 7.

and the fear of the LORD. His delight shall be in the fear of the LORD. He shall not judge by what his eyes see, or decide by what his ears hear; but with righteousness he shall judge the poor, and decide with equity for the meek of the earth; he shall strike the earth with the rod of his mouth, and with the breath of his lips he shall kill the wicked. Righteousness shall be the belt around his waist, and faithfulness the belt around his loins. The wolf shall live with the lamb, the leopard shall lie down with the kid, the calf and the lion and the fatling together, and a little child shall lead them. The cow and the bear shall graze, their young shall lie down together; and the lion shall eat straw like the ox. The nursing child shall play over the hole of the asp, and the weaned child shall put its hand on the adder's den. They will not hurt or destroy on all my holy mountain; for the earth will be full of the knowledge of the LORD as the waters cover the sea.

Figure 8. "The Peaceable Kingdom", The Illuminated Bible, *New York 1846.*

Another potent eschatological vision is the New Jerusalem. To get to that Edenic harmony, the present world will be transformed by healing.

Revelation 21:2–4. And I saw the holy city, the new Jerusalem, coming down out of heaven from God, prepared as a bride adorned for her husband. And I heard a loud voice from the throne saying, "See, the home of God is among mortals. He will dwell with them; they will be his peoples, and God himself will be with them; he will wipe every tear from their eyes. Death will be no more; mourning and crying and pain will be no more, for the first things have passed away."

Figure 9: "The New Jerusalem," Gustav Dore, 1865 (source unknown).

This is a vision of completion, healing, fulfillment. With the new creation, the old creation will attain its full quidity. Omega will redefine alpha.

The Sacrament of the Altar anticipates the eschatological omega. The Eucharist unites creation with redemption, present with future. Liturgically, we sing of the bread and wine as a "foretaste of the feast to come." This leads Rosemary Radford Ruether to write, "this redemptive communion in the whole creation points forward to its eschatological completion. Communion in the body and blood of Christ is a foretaste of that great communion, when the Word of God, who brought both the whole creation and descended to become part of creation, gathers up the whole creation into its culminating union with God. Then God will be 'all in all'."[32]

According to our retroactive ontology, God's promise to fulfill this vision constitutes fundamental reality. Past and present gain their respective definitions from this destiny. Who we are today is on the way, so to speak, to the reality promised by God. When we live in hope, we anticipate this omega. Anticipation informs our ethics, drives our ethics.

Seven Middle Axioms

These biblical visions are symbolic. We do not take them literally, yet we take them seriously. The truth conveyed through these symbols is that the actual world we see every day is not all there is. Reality is much more. Biblical symbols drive the mind beyond itself, to a contemplation of divine transcendence accompanied by a divine presence.

We must interpret the symbols by abstracting or constructing theological concepts such as this: God's eschatological promise includes a healing harmony that binds together our inner self, other selves, the natural world, and the divine reality. All these together constitute the whole of reality subject to divine redemption. Such a theological interpretation places us in a position of formulating middle axioms that can bridge the gap between an encompassing eschatological vision, on the one hand, and our ethical responsibility in a historical world that is yet far from embodying this vision, on the other. The comprehensive vision draws us ahead, inspiring partial yet proleptically significant transformatory enterprises.

For some years now I have been road testing various sets of middle axioms that draw out ethical and even ecological guidelines from proleptic hope.[33] Each begins with the Greek prefix, *pro*, connoting anticipation of actualities yet to come. Let me repeat them in slightly revised form here.

1. Project a vision of the coming new order that is
   (1) organized as a single, worldwide, planetary society;
   (2) untied in devotion to the will of God;

---

[32] Rosemary Radford Ruether, "Ecological Theology: Roots in Tradition, Liturgical and Ethical Practice for Today," *Dialog*, 42:3 (Fall 2003) 233.

[33] These provolutionary principles in various expressionss have appeared in Ted Peters, *Futures – Human and Divine* (Louisville: Westminster John Knox Press, 1977) 170–181; *Fear, Faith, and the Future* (Minneapolis: Augsburg Press, 1980) 113–120; and *GOD – The World's Future* (Minneapolis: Fortress Press, rev. ed., 2000) 378–392.

(3) sustainable within the biological carrying capacity of the planet and harmonized with the principles of the ecosphere;
(4) organized politically so as to preserve the just rights and voluntary contributions of all individuals;
(5) organized economically so as to guarantee the basic survival needs of each person;
(6) organized socially so that dignity and freedom are respected and protected in every quarter;
(7) dedicated to advancing the quality of life on behalf of future generations.

2. Promote a sense of global community.
3. Provide for posterity.
4. Protect – and perhaps promote – human dignity.
5. Proffer the distinction between needs and wants.
6. Propose alliances between Christians and non-Christians of good will and common purpose.
7. Profess faith in God's future even in the face of disappointment.

3. The Prophetic Proviso *Thesis:*
The eschatological vision needs an accompanying prophetic proviso – that is, inspiration for transformation accompanied by a precautionary warning against immanentizing the transcendent.

I have been suggesting that futurists and ecologists are akin to prophets. They look somewhat like the biblical prophets because they speak of what will happen in the future; and they challenge the present generation to wake up and take concerted moral action. They foresee the necessary connection between planet sustainability and social justice. They issue calls to our consciences and urge responsible action.

Larry Rasmussen is one of the more passionate theologians who places Christian ethics into partnership with secular ecological ethics. He connects the biblical prophets with the hope for liberation in what he calls "prophetic-liberative practices." He writes, "Hope and redemption (peace and abundance) is the prophets' baseline. And hope is always hope for the redemption of creation as a whole, the people and the land together. This is redemption as the liberation of all life, from the cell to the community, a struggle inclusive of the poor, the weak, the marginalized, the diseased and the disfigured. It is, not least, liberation of exploited and exhausted nature."[34]

---

[34] Larry Rasmussen, "Drilling in the Cathedral," *Dialog* 42:3 (Fall 2003) 215.

The Eschatological Problem: Distinguishing the Imminent from the Immanent

God's eschatological future is *imminent*. It is not *immanent*. Even though God draws the world moment by moment beyond its present state of being toward creative transformation, the ultimate transformation remains future. It remains transcendent. It can be proleptically incarnated, but not exhaustively incarnated let alone manipulated or controlled.

The eschatological problem asks: how do we get there from here? How can we get from our present world order replete with competition between predator and prey or competition between peace and war to the world order envisioned by the Peaceable Kingdom or the New Jerusalem? The answer is clear: only by God's grace and power. We can and should live today out of the vision of what is to come; but arriving there will not be our own doing. Our own doing has to do with our present reality, subject to transformation in light of our transcendental vision.

This is where the element of prophetic judgment enters Christian theology. On the one hand, the prophetic accent draws us forward, toward omega. On the other hand, the prophetic voice renders judgment against premature or delusional thinking that we can attain it fully through concerted political or economic or even ecclesial action.

A prophetic caution is called for. It is a caution against over-identifying the vision of the eschatological omega with our own efforts at historical transformation. On the one hand, we need to unite communally in the single task of incarnating our vision of a better world. On the other hand, it is just this promise of transformation that places in the hands of ruthless leaders – whether national or international leaders – the power they need to take totalitarian control. The core of the danger is found in the delusional immanentizing of the transcendent.

A prophetic shock to theology came in the 1930s in response to the new hegemonies of Marxism and fascism, especially Nazism in central Europe. The Barmen Declaration of 1934 took a stand against Adolf Hitler: "We reject the false doctrine that ... the Church could ... allow itself to be given special leaders [*Führer*] vested with ruling authority."[35]

Writing in Vienna in 1938, Eric Voegelin unmasked the disguise of secular politics to show that the modern state, especially the totalitarian state, belongs "to the sphere of the religious."[36] This politi-

[35] The Barmen Declaration, § 4. See: "II. Barmer Theologische Erklärung" by Carsten Nicolaisen in *Religion in Geschichte und Gegenwart*, ed. by Hans Dieter Betz/ Don S. Browning/Bernd Janowski/Eberhard Jüngel (8 Volumes: Tübingen: Mohr Siebeck, [4]1998–2005) 1:1111–1115.

[36] Eric Voegelin, "Modernity Without Restraint: The Political Religions," *The Collected Works of Eric Voegelin* (34 Volumes: Baton Rouge/London: Louisiana State University Press, 1990–1995) 5:28.

cal philosopher distinguished between the trancencdent and the immanent when distinguishing the spiritual from the political spheres, even though both are religious. "The spiritual religions, which find the *realissimum* in the Ground of the world (*Weltgrund*), should be called trans-worldly religions (*überweltliche Religionen*); and all others, i.e. those that find the divine in subcontents of the world, should be called inner-worldly religions (*innerweltliche Religionen*)."[37]

With this distinction in hand, Voegelin could then describe the role of myth in modern society. Ancient religions organized their world view through myths. Modern political entities also rely on myths, disguised myths. "The 'myth' is created purposely to bind the masses emotionally and to arouse in them the politically effective expectation of salvation. Since the myth cannot legitimize itself thorough transcendent revelation or stand up to scientific criticism, a new concept of truth is developed … only that is true which promotes the existence of the organically closed, inner-worldly national community."[38]

Alluding to Adolf Hitler, Voegelin compared the German *Führer* to transcendentally grounded religious authorities of old. "The Führer is the point where the spriit of the people breaks into historical reality; the inner-worldly god speaks to the Führer in the same way the transcendent God speaks to Abraham, and the Führer transforms God's words into commands for his immediate followers and for the people."[39]

What we should learn from this, I believe, is that when projecting our vision of God's eschatological omega we should avoid the trappings of political mythology, of falsely deifying an ideology, party, or leader. Paul Tillich, writing in Germany in 1934, two years after Hitler took power, alerted us to the same danger. "The concentration of all spheres of life within the unlimited authority of the national state is possible only when founded upon a world view which has the inherent power of encompassing man's entire being and driving him on to uncondtitional self-surrender. Such a world view is religious in character and finds expression in a myth. The more unconditional and more inclusive the claims of the state are, the more fundamental and powerful must be the myth which is the foundation of such claims."[40]

The key insight is that justification for totalitarianism is borrowed from disguised transcendental claims. These transcendental claims can take immanentized eschatological form. The very phrase, "The Third Reich" (*das*

[37] Ibid., 32–33.

[38] Ibid., 62.

[39] Ibid., 65.

[40] Paul Tillich, "The Totalitarian State and the Claims of the Church," in *Main Works/Hauptwerke*, Carl Heinz Ratschow, editor in chief (6 Volumes: Berlin/New York: De Gruyter, 1989) 3:430.

Dritte Reich), for example, is a mythological phrase which promises a millennial reign of one thousand years of German hegemony and prosperity. Tillich and Voeglein along with other twentieth century theologians criticized such thinking, describing it as dangerously utopian.

Now, let us apply this theological reaction to the question at hand. Eric Voegelin feared "immanentist eschatology," where the transcendent eschaton is presumed to have come under human direction and human control. This is a delusion, of course; yet it is a tempting delusion. Voegelin used the term "gnosticism" to describe this attempt to domesticate the otherwise illusive transcendent either through knowledge or through political power. "The attempt at immanentizing the meaning of existence is fundamentally an attempt at bringing our knowledge of transcendence into a firmer grip than the *cognitio fidei*, the cognition of faith, will afford; and gnostic experiences offer this firmer grip in so far as they are an expansion of the soul to the point where God is drawn into the existence of man."[41]

The doctrine of progress in western culture provides a salient example of the immanentization of eschatology, the domesticating of divine purpose. Those who would ascribe obliquely divine power and authority to themselves place themselves in the position of guiding if not controlling progress. Because progress has become secularism's sacred, the magnets of progress become the high priests of the world economy if not our nation states. "Totalitarianism, defined as the existential rule of gnostic activists, is the end form of progressive civilization."[42] By "end form" here, Voegelin did not mean only the epitome; he also meant decline and self-destruction.

In summary, the prophetic proviso cautions us against surrendering individual human dignity to totalitarian rule, and against collapsing our eschatological vision into the economic or even ecological ideology of an attractive political party. It cautions us against sacralizing the natural world or, worse, sacralizing the political party claiming to lead us into a final eco-salvation.

The Prophetic Proviso

For Paul Tillich, the prophetic accent within the Christian faith, though perpetually available throughout church history, is best ar-

---

[41] Eric Voegelin, *The New Science of Politics*, in *The Collected Works of Eric Voegelin* (34 Volumes: Baton Rouge/London: Louisiana State University Press, 1990–1995) 5:189.

[42] Ibid., 195.

ticulated by the theologians of the Protestant Reformation. What is essential to the prophetic accent is the repeated reminder that God is transcendent. God cannot be captured, contained, or constrained by human thought, ritual, or political institution. Even in revelation, God remains mysterious, incomprehensible. To confuse our finite reality with the divine is to fail to recognize the infinite qualitative difference between the eternal God and the created order. To pretend to have carnally or temporally institutionalized the divine reality is idolatry. Prophetic judgment protects us from idolatry.

In our own modern era, Tillich calls this prophetic reminder the *Protestant Principle.* This principle renders a judgment "against the identification of our ultimate concern with any creation of the church," or any other human institution, for that matter.[43] "The divinity of the Divine, the ultimacy of the Ultimate, the unconditional character of the Unconditioned. That is the point where the Reformation starts."[44] Let me emphasize: this is the point at which the Reformation "starts." What Tillich is pointing us to cannot be regularized, institutionalized. Rather, it is a dialectical principle. Even Protestants can lose the prophetic dialectic if they are not vigilant. "The more Protestantism becomes actualized the more the prophetic spirit becomes embodied in an organization; and this organization shows anti-prophetic traits."[45] The prophet does not establish community; the prophet provides the word of transcendental direction and divine judgment to the community.

When it comes to the relationship of the eschatological kingdom of God to temporal political orders, a certain confusion exists already within our inherited Christian symbols. "We have in the Book of Revelation two symbols for the end – end in the sense of 'finish' and at the same time of 'the aim of history'. The one is the eschatology which says the last things will come *after* history, in something beyond time and space, in the fulfilled kingdom of God. At the same time, we have in the same book the idea of a rule of Christ, symbolized as ' thousand years', which will be *in* history. The conflict of these two ideas ... has moved our Church history and has permanently produced the problem of the prophetic element in the history of the Church."[46] The chiliastic or millennialist sub-tradition within Christianity risks compromising the prophetic accent of eschatology with idolatrous claims supporting an

---

[43] Paul Tillich, *Systematic Theology* (3 Volumes: Chicago: University of Chicago Press, 1951–1963) I:37.

[44] Paul Tillich, "The Recovery of the Prophetic Tradition in the Reformation," in *Main Works/Hauptwerke*, Carl Heinz Ratschow, editor in chief (6 Volumes: Berlin/New York: De Gruyter, 1989) 6:326.

[45] Ibid., 352.

[46] Ibid., 320.

immanentized political program. The prophetic voice – now in the form of the Protestant Principle – must be raised in judgment against the absolutizing of any political program, none of which can be identical to the eschatological vision of the Peaceable Kingdom or the New Jerusalem.

The subtlety of the dialectic includes the following observation. What we consider to be our temporal hopes – hopes for technological progress, political revolution, advances in human insight, advances in spiritually united community, life fulfillment – are implicitly rooted in the religious dimension of human consciousness.[47] They derive from a primordial trust in what transcends our ordinary circumstance. So the task of the eschatological prophet is both to articulate our implicit trust in the divine draw toward transformatory fulfillment while, at the same time, declaring judgment regarding the difference between that which transcends history and what is immanent to history.

Wolfhart Pannenberg refers to this prophetic accent – both the future orientation as well as the caution against immanentizing – with the term *provisionality.* The church should play the role of the prophetic conscience to the state, says Pannenberg. In light of the eschatological promise, "all human ordering of social life in this world is provisional, far distant from the definitive actualizing of God's kingdom by God himself. That there is this great difference is the reason why the church has to exist as a special society alongside the state, even though the theme of the political order has in fact the content that will find definitive fulfillment in God's kingdom, i. e., human life together in justice and peace."[48]

The Life of Beatitude

Retroactive ontology grounds the Life of Beatitude. Anticipation of omega incarnates the future ahead of time. Our life of hope is based upon God's promise to provide eschatological confirmation of what we now anticipate.

The ethic appropriate to the structure of the Life of Beatitude takes the form of a proleptic ethic. A proleptic form of ethics begins with a vision of God's eschatological future and then formulates middle axioms to guide actions on behalf of historical transformation. Because we know that the historical world is undergoing change, a proleptic approach to ethics does not seek the good in the past or in the present

[47] Paul Tillich, "Die christliche Hoffnung und ihre Wirkung in der Welt," ibid., 424.

[48] Wolfhart Pannenberg, *Systematic Theology,* tr. by Geoffrey W. Bromily (3 Volumes: Grand Rapids MI: Wm. B. Eerdmans, 1991–1998). III:56.

state of being, but rather in future transformation. It seeks the good not in a *past genesis* but in a *future epigenesis*.

Carl Braaten has referred to what I think of as beatitudinal living in terms of "exchatopraxis." He once wrote, "in proleptic ethics it may truly be said that the end justifies the means, because the end is proleptically present and operative beforehand, rehearsing the qualities of the eschatological kingdom – peace, love, joy, freedom, equality, unity – in the course of history's forward movement."[49]

Our anticipation of omega manifests itself in proleptic ethics, a prophetic dialectic of vision and transformation balanced by caution and judgment. The divine promise is that a new creation is imminent; the Kingdom of God is at hand. This provides the source of our vision of a transformed future. Our ethical mandate is to live in hope today out of the power of tomorrow's reality. Yet, the prophetic proviso reminds us that God's consummate future will require an act on God's part, an ultimate transformation that only God can deliver. In the meantime, when we treat each person regardless of innate capacity with dignity, we participate proleptically in the eschatological consummation yet to come in its fullness.

[49] Carl E. Braaten, *Eschatology and Ethics* (Minneapolis: Augsburg, 1974) 121.

Index

Name Index

Subject Index